Practical Chemistry *for*

E. N. Ramsden

BSc, PhD, DPhil

Basil Blackwell

First published 1987
Reprinted 1987 (with corrections)

Published by Basil Blackwell Ltd
108 Cowley Road
Oxford OX4 1JF
England

Illustrated by Anne Langford

Typeset in 13 on 15pt Melior
by Columns of Reading

Printed and bound in Great Britain

Contents

Acknowledgements *viii*

Preface *ix*

Introduction Safety in the laboratory x

Exercise Spotting the hazards (poster issued by the Royal Society for the Prevention of Accidents) *xi*

Topic 1 Solids, liquids and solutions *1*

Experiment 1.1 Aftershave: some market research *1*
Experiment 1.2 Growing large crystals of alum (aluminium potassium sulphate) *3*
Experiment 1.3 Plotting a solubility curve for potassium chloride *5*
Experiment 1.4 Removing graffiti *7*

Topic 2 Methods of separating mixtures *9*

Experiment 2.1 Obtaining pure salt from rock salt *9*
Experiment 2.2 Separating sodium chloride, lead(II) chloride and sand *11*
Experiment 2.3 Separating black ink into a solid and a liquid *12*
Experiment 2.4 Separating ethanol and water by distillation *13*
Experiment 2.5 Is the pigment in black ink a pure substance? *15*
Experiment 2.6 Using chromatography to identify a dye *17*
Experiment 2.7 Chromatography on petals and leaves *19*
Experiment 2.8 Column chromatography on food colourings *21*
Experiment 2.9 Plotting a cooling curve for a molten solid *23*

Topic 3 Elements and compounds *25*

Experiment 3.1 What happens when iodine meets hot iron? *25*
Experiment 3.2 Mixing tin and iodine *26*
Experiment 3.3 What happens when sulphur meets hot copper? *27*
Experiment 3.4 Heating a mixture of iron and sulphur *28*

Topic 4 Radioactivity *31*

Exercise 4.1 Plotting radioactivity against time *31*

Topic 5 Electrochemistry *33*

Experiment 5.1 Finding out which substances conduct electricity *33*
Experiment 5.2 What happens when solutions conduct electricity? *35*
Experiment 5.3 Why does electricity make some substances move? *36*
Experiment 5.4 Combining zinc and iodine and passing a direct current through the product *37*
Experiment 5.5 Electrolysis of copper(II) sulphate solution with copper electrodes *38*
Experiment 5.6 A test circuit *39*
Experiment 5.7 Copper plating (and zinc plating) *40*
Experiment 5.8 The lead–acid accumulator *41*
Experiment 5.9 Anodising and dyeing aluminium *42*
Experiment 5.10 Writing with electricity *44*

Topic 6 Formulas *47*

Activity 6.1 Formula cards *47*

Topic 7 Acids, bases and salts *49*

Experiment 7.1 Reactions of acids *49*
Experiment 7.2 Reactions of bases *51*
Experiment 7.3 What do you obtain when you neutralise vinegar with lime? *53*
Experiment 7.4 Indicators *54*
Experiment 7.5 Making zinc sulphate crystals *56*
Experiment 7.6 Making crystals of copper(II) sulphate-5-water *57*
Experiment 7.7 Heating crystals of copper(II) sulphate-5-water *58*
Experiment 7.8 Preparation of sodium sulphate crystals *60*
Experiment 7.9 Preparation of insoluble salts *62*
Preparation of lead(II) chromate *62*
Preparation of lead(II) iodide *62*
A chemical garden *62*
Experiment 7.10 Comparing indigestion tablets *63*
Experiment 7.11 A long experiment to do at home *64*

Topic 8 Air *65*

Experiment 8.1 Burning elements in oxygen *65*
Experiment 8.2 Under what conditions does iron rust? *68*
Experiment 8.3 What percentage of the air is used in rusting? *70*
Experiment 8.4 Does sulphur dioxide affect plant growth? *71*
Experiment 8.5 Some reactions of hydrogen peroxide *72*

Topic 9 Water *75*

Experiment 9.1 Which ions make water hard? *75*
Experiment 9.2 Methods of softening tap water *76*

Experiment 9.3 Methods for getting rid of a film of oil on water *77*
Experiment 9.4 Comparing soaps with detergents *78*
Experiment 9.5 Comparing different detergents *80*
Experiment 9.6 Which work better: enzyme detergents or non-enzyme detergents? *82*
Experiment 9.7 Should you use the same detergent for every type of fabric? *83*
Experiment 9.8 Matching the cleaner to the stain *84*
Experiment 9.9 A soil problem *84*

Topic 10 Metals *85*

Experiment 10.1 Reactions of metals with water and with dilute acids *85*
Experiment 10.2 Studying the reaction of calcium with water *87*
Experiment 10.3 Studying the reaction of magnesium with steam *88*
Experiment 10.4 Studying the action of steam on iron *89*
Experiment 10.5 Putting metals in order of reactivity *91*
Experiment 10.6 Cleaning the silver — an experiment to do at home *93*
Experiment 10.7 What happens when you heat carbon with a metal oxide? *94*
Experiment 10.8 Electricity from metals *95*
Experiment 10.9 Focus on iron *96*
Experiment 10.10 Investigating rusting *97*
Experiment 10.11 Corrosion of pairs of metals *99*
Experiment 10.12 Focus on aluminium *101*
Experiment 10.13 Removing the oxide film from aluminium *102*

Topic 11 Formulas and equations *105*

Experiment 11.1 Finding the formula of magnesium oxide *105*
Experiment 11.2 Finding the formula of copper oxide *106*
Experiment 11.3 Finding the balanced equation for a reaction *108*
Experiment 11.4 Finding the formula of magnesium sulphate crystals *109*

Topic 12 Carbonates and carbon dioxide *111*

Experiment 12.1 Heating calcium carbonate *111*
Experiment 12.2 Making mortar *113*
Experiment 12.3 Investigating carbonates *114*
Heating carbonates *114*
Action of acid on carbonates *114*
Experiment 12.4 Preparation of carbon dioxide *116*
Experiment 12.5 Fire extinguishers *119*
Experiment 12.6 How easily do fabrics burn? *121*
Experiment 12.7 Investigating shells *122*

Topic 13 Sulphur and its compounds *125*

Experiment 13.1 Crystals of sulphur *125*
Experiment 13.2 Finding out what is formed when sulphur burns *127*
Experiment 13.3 Alum: a useful sulphate *129*
Experiment 13.4 Sodium hydrogensulphite: some uses *131*
Experiment 13.5 Invisible writing *132*

Topic 14 Nitrogen compounds *133*

Experiment 14.1 Heating protein foods with soda–lime *133*
Experiment 14.2 Making the fertiliser ammonium sulphate *135*
Experiment 14.3 Making ammonia *136*
Experiment 14.4 Making the fertiliser ammonium phosphate *138*
Experiment 14.5 Testing the fertilisers that you have made *139*
Experiment 14.6 Two ammonium chloride problems *140*

Topic 15 Halogens *141*

Experiment 15.1 Investigating some reactions between halogens *141*
Experiment 15.2 Iodine gets another chance *143*
Experiment 15.3 Comparing household bleaches *144*

Topic 16 Hydrocarbons *145*

Experiment 16.1 Obtaining fractions from crude oil *145*
Experiment 16.2 Experiments with alkanes *147*

Topic 17 Alkenes and alcohols *149*

Experiment 17.1 Heating liquid paraffin *149*
Experiment 17.2 Investigating unsaturation in fats and oils *150*
Experiment 17.3 Fermenting apple juice *151*
Experiment 17.4 Studying some reactions of ethanol *153*
Experiment 17.5 Making polymers *154*
Experiment 17.6 A key to identifying fabrics *156*

Topic 18 Heat changes in chemical reactions *159*

Experiment 18.1 Some reactions which involve heat changes *159*
Experiment 18.2 Thermometric titration *161*
Experiment 18.3 Observing temperature changes when copper is displaced from copper(II) sulphate solution *163*
Experiment 18.4 Comparing different fuels *165*

Topic 19 Speeds of reactions *167*

Experiment 19.1 Does the speed at which a solid reacts depend on the size of the solid particles? *167*

Experiment 19.2 How does the concentration of a reactant affect the time taken for the completion of a chemical reaction? *168*
Experiment 19.3 Investigating the effect of concentration on the speed of the reaction between acid and thiosulphate *170*
Experiment 19.4 Investigating the effect of temperature on the speed of a chemical reaction *172*
Experiment 19.5 The effect of light on a chemical reaction *174*
Experiment 19.6 Catalysis *175*

Appendix List of chemicals and apparatus needed for each experiment *177*

Acknowledgements

During the term that I spent preparing this book, I enjoyed a teacher studentship at St Hugh's College, Oxford. I am deeply grateful to the Principal and Fellows of the College for the stimulating and enriching time which I spent with them and for the opportunity to work undistracted by the demands of school.

I would like to thank Professor J. B. Goodenough and Dr J. C. Green for inviting me to use the facilities of the Inorganic Chemistry Laboratory at Oxford while I was working through the experiments. It was a pleasure to work in such a hospitable department. I thank Mr D. Chapple and Mr K. Waters especially for the assistance I received from them.

I thank the staff of Basil Blackwell for helpful discussions during the writing of the book and for their careful work in preparing it for publication.

I thank my family for their support during the whole of the project.

E. N. Ramsden, 1987

Preface

Practical Chemistry for GCSE is a support to my text *Chemistry for GCSE* (Basil Blackwell, 1986). It is a collection of experiments which I have selected for different reasons. Among reasons for doing practical work are the need to get the feel of the subject, the opportunity of developing practical skills and the value of developing an investigative, problem-solving approach to science. Some of the experiments are included simply because they lay the foundation for material covered in the text. Others are included because of their visual appeal. In some experiments, the factual content is less important than the scientific process involved. There are suggestions for experiments and investigations in which pupils are invited to take a major part in the design of the experiment. Some pupils may need guidance from the teacher, but others will be able to manage with the help of the hints that are offered. Since some areas of study lend themselves more readily than others to safe experiments, some topics are longer than others. The book is not a prescription for practical chemistry: teachers may need to repeat some experiments which I have omitted on the grounds that they will have been covered in the pre-certificate years, and naturally there are many interesting experiments which teachers may wish to add.

A large number of these experiments can easily be converted by teachers into GCSE assessment exercises, but because assessment schemes vary so widely between examining groups I have not attempted to introduce mark schemes.

E. N. Ramsden, 1987

Introduction
Safety in the laboratory

There are certain rules for working in chemistry laboratories.

1 Work only under the supervision of your teacher. If you have ideas for experiments of your own, check them with your teacher first.

2 Concentrate on your own experiment. Do not interfere in another pupil's experiment.

3 Do not move around more than is necessary. Do not run.

4 **Wear safety glasses**. If you have long hair, tie it back.

5 When heating a chemical in a test-tube, make sure that you are not pointing it at your neighbour or at yourself.

6 Use clean apparatus. Tidy up after your work. Put solid waste into a bin, and not into a sink. Let hot objects cool before you put them away.

7 Do not taste chemicals. Smell gases cautiously.

8 If you have an accident, a burn, a cut or a splash of some chemical, wash with plenty of cold water. Tell your teacher at once.

Exercise   Spotting the hazards

Study the illustration. List all the situations which could lead to accidents.

Poster issued by the Royal Society for the Prevention of Accidents

TOPIC 1 Solids, liquids and solutions

Experiment 1.1
Aftershave: some market research

Note: Aftershave lotions are used to cool the skin after shaving. They change state from liquid to gas. The change of state produces a cooling effect. You can experiment to find out which brand of aftershave produces the largest cooling effect.

1 Make an aftershave lotion. Shake together thoroughly the following:
8 cm^3 of methylated spirit,
1 cm^3 of glycerol,
4 cm^3 of water.

2 Set up the apparatus shown in Fig. 1.1. The quantity of cotton wool should be just enough to absorb 2 cm^3 of liquid. Read the thermometer, and write down the temperature.

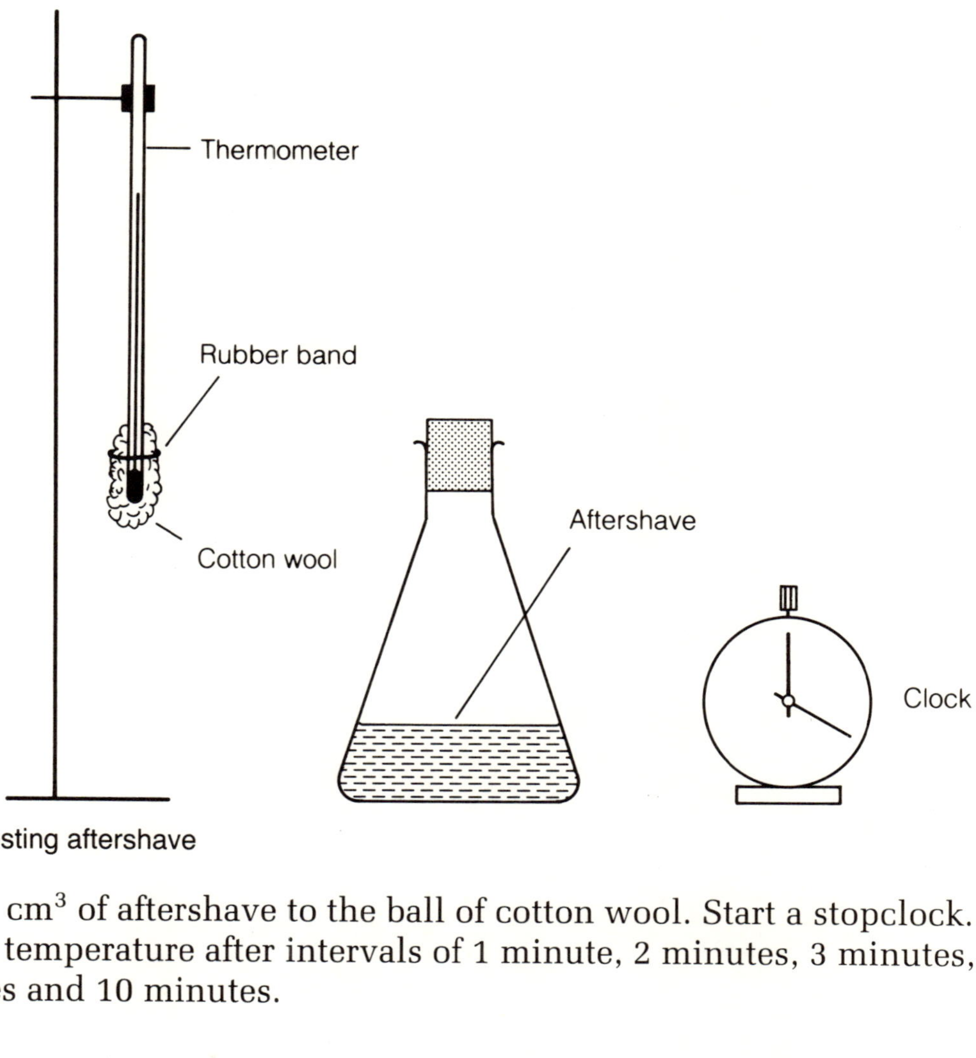

Fig. 1.1 Testing aftershave

3 Add 2 cm^3 of aftershave to the ball of cotton wool. Start a stopclock. Note the temperature after intervals of 1 minute, 2 minutes, 3 minutes, 5 minutes and 10 minutes.

4 Now repeat the experiment, using water. This time, add 2 cm^3 of water, instead of aftershave, to the cotton wool. Record the temperature after intervals of 1, 2, 3, 5 and 10 minutes. How well does water compare with aftershave?

5 For other experiments, use commercial brands of aftershave which have been brought from home. You can also make more varieties of your own by varying the proportions of the three ingredients given in step **1**.

6 Why does the skin become warm during shaving?
How do aftershave lotions cool the skin?
Why do they work better than water?
Which of the aftershaves tested do you recommend?

A small project

In this experiment, you have compared the rates at which different liquids evaporate. Now focus your attention on one liquid, e.g. water. There are various factors which can change the speed with which water evaporates. Take 10 cm^3 of water in a 100 cm^3 beaker. Suggest ways in which you could make it evaporate faster. Try out your ideas.

Experiment 1.2
Growing large crystals of alum (aluminium potassium sulphate)

1 First, make a saturated solution of aluminium sulphate. Take 100 cm^3 of water in a 250 cm^3 beaker. Add a spatula measure of aluminium sulphate and stir with a glass rod (see Fig. 1.2a). Continue adding aluminium sulphate, stirring in a spatula measure at a time, until no more will dissolve.

2 Filter the saturated solution (see Fig. 1.2b).

3 Take 100 cm^3 of water in a second beaker. Make a saturated solution of potassium sulphate. Filter it.

4 Mix the two solutions.

5 Evaporate the combined solution over a water bath until crystallisation just begins (see Fig. 1.2c).

6 Add a little water to dissolve the crystals that have formed. Allow the solution to cool.

7 When the temperature has dropped to 50 °C, suspend a thread in the solution. Crystals will soon form on it. Choose the one that has the best shape. Remove the others.

8 Leave the thread running through the chosen crystal. Hang it in the solution (see Fig. 1.2d).

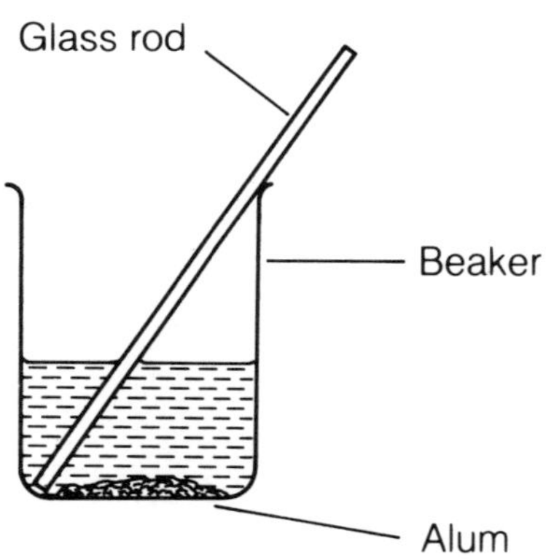

Fig. 1.2(a) Making a solution

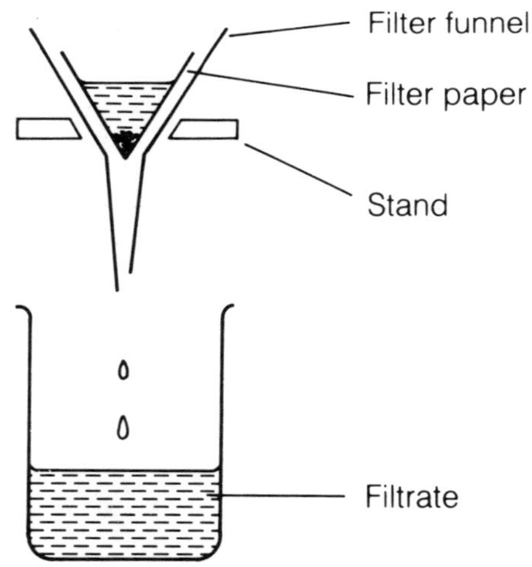

Fig. 1.2(b) Filtering

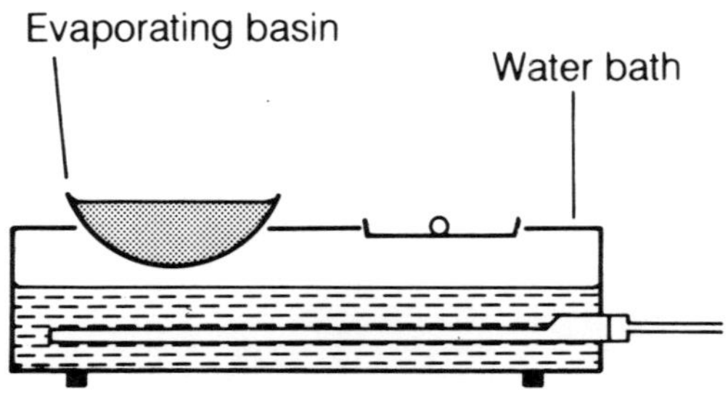

Fig. 1.2(c) Evaporating on a water bath

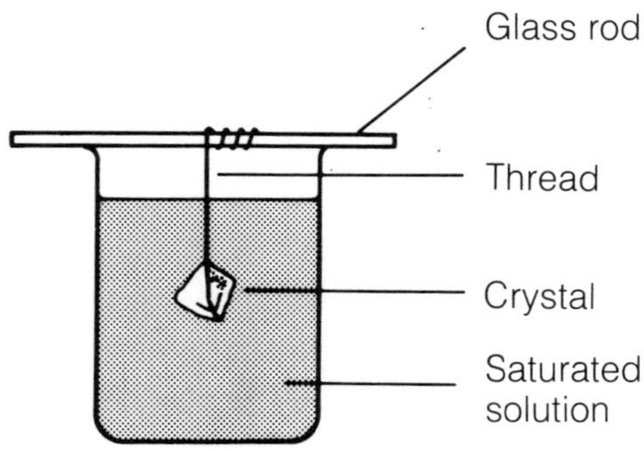

Fig. 1.2(d) Growing a crystal

9 In a week, the crystal will have grown. If you want it to go on growing, you will have to give it some more solution to grow in. Prepare some more saturated alum solution. Cool the solution. Hang the crystal in it. If the solution is not saturated, some of the crystal will dissolve. If you have managed things just right, your crystal will grow and grow.

10 Extension work: Hang the crystal in a saturated solution of chromium(III) potassium sulphate. (This is made from potassium sulphate and chromium(III) sulphate in the method described in steps **1–6**.) When it stops growing, return the crystal to a saturated aluminium potassium sulphate (alum) solution. You can get an interesting result by alternating the crystal between the solutions of the two different alums.

Experiment 1.3
Plotting a solubility curve for potassium chloride

Note: The solubility of a substance is the mass of the substance that will saturate 100 g of a solvent at a certain temperature. A saturated solution is one that cannot dissolve any more of the solute at that temperature.

1 Weigh 10.0 g of potassium chloride.

2 Tip the salt into a boiling tube. Add 20 cm^3 of water.

3 Place the boiling tube in a beaker of water on a tripod and gauze (see Fig. 1.3a). Heat the beaker and stir the solution until all the potassium chloride has dissolved.

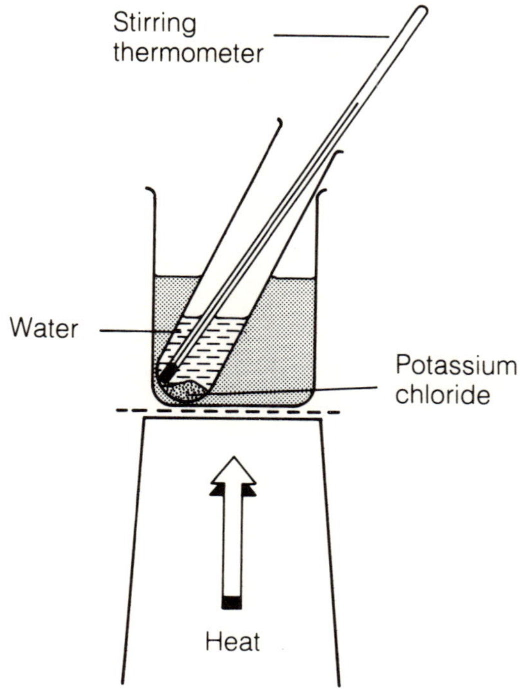

Fig. 1.3(a) Heating

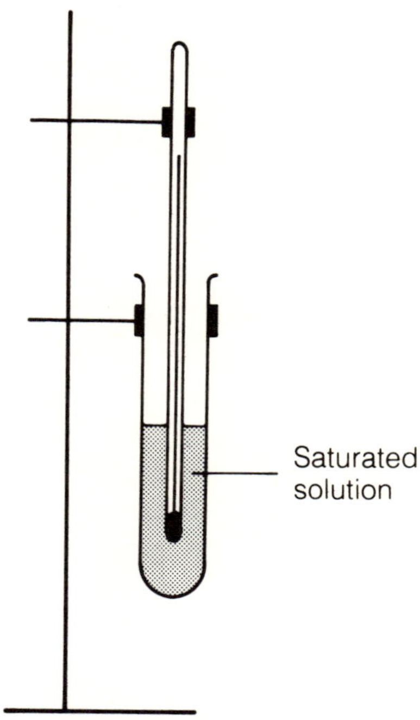

Fig. 1.3(b) Cooling

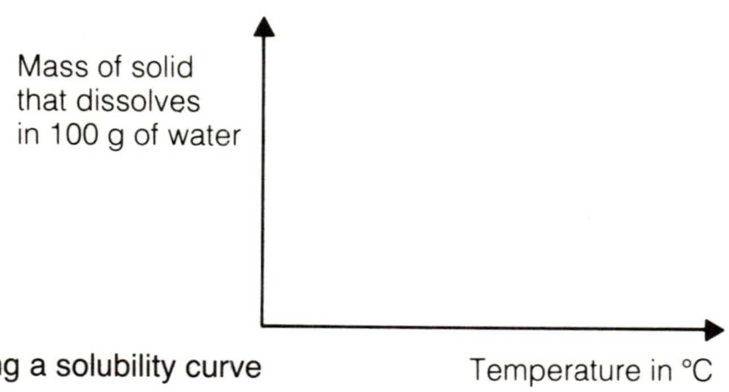

Fig. 1.3(c) Plotting a solubility curve

4 Take the boiling tube out of the hot water. Allow the solution to cool. Stir constantly. Note the temperature at which crystals first appear (Fig. 1.3b).

5 Add 2.5 cm^3 of water. Repeat steps **3** and **4**. Find the new temperature at which crystals appear.

6 Add another 2.5 cm^3 of water, and repeat. Obtain six results. You will have to cool the beaker with ice–water in the fifth and sixth experiments.

7 Copy this table, and enter your results.

Table: Solubility of potassium chloride

Result	*Total volume of water in* cm^3	*Temperature at which crystals form in* °C	*Solubility in* g/100 cm^3 *of water*
1	20.0		50.0
2	22.5		44.5
3			
4			
5			
6			

8 Plot your results on graph paper (see Fig. 1.3c). Other solutes you can use are sodium chloride, potassium chlorate(V), copper(II) sulphate and sodium nitrate.

Experiment 1.4
Removing graffiti

1 Assemble stoppered bottles containing small quantities of various solvents. Label each bottle. Some solvents you can use are water, soap solution, detergent solution, ammonia solution, ethanol, methanol, propanone, ethyl ethanoate, trichloroethane and dichloromethane.

2 Find some graffiti. Experiment to find out which solvents are good for removing graffiti. Make a table of your results (Table A).

3 Calculate the price of 100 cm^3 of each solvent. Include the prices of the solvents in Table A.

Table A Removing graffiti with different solvents

Solvent	*Type of graffiti*	*Type of surface*	*How well did the solvent work?*	*Price of 100 cm^3 of solvent*
	Ballpoint pen *Felt-tip pen* *etc.*			

4 Compare the effectiveness of the solvents with their prices. Which solvent or solvents would you recommend?

5 Now do some tests using one solvent only. Find out how good it is at removing graffiti from different surfaces, e.g. varnished wood, plain wood, painted wood, emulsion paint and brick. Repeat the tests on a second solvent. Tabulate your results (Table B).

Table B Removing graffiti from different surfaces

	Varnished wood	*Unvarnished wood*	*Painted wood*	*Emulsion paint*	*Others*
Solvent A					
Solvent B					

6 As a result of your tests, say what types of graffiti are most difficult to remove. Try to find out how much your school spends on removing graffiti.

TOPIC 2 Methods of separating mixtures

Experiment 2.1
Obtaining pure salt from rock salt

You are asked to obtain pure salt from rock salt.

The substance we normally call *salt* is sodium chloride. It is mined as *rock salt*, a mixture of salt, rock and sand. You can use the solubility of salt in water to separate pure salt from the other substances present in rock salt.

You are supplied with rock salt, pestle, mortar, beaker, glass rod, filter funnel, filter paper, tripod and gauze, glass rod, evaporating basin, water bath and other normal laboratory apparatus (Fig. 2.1).

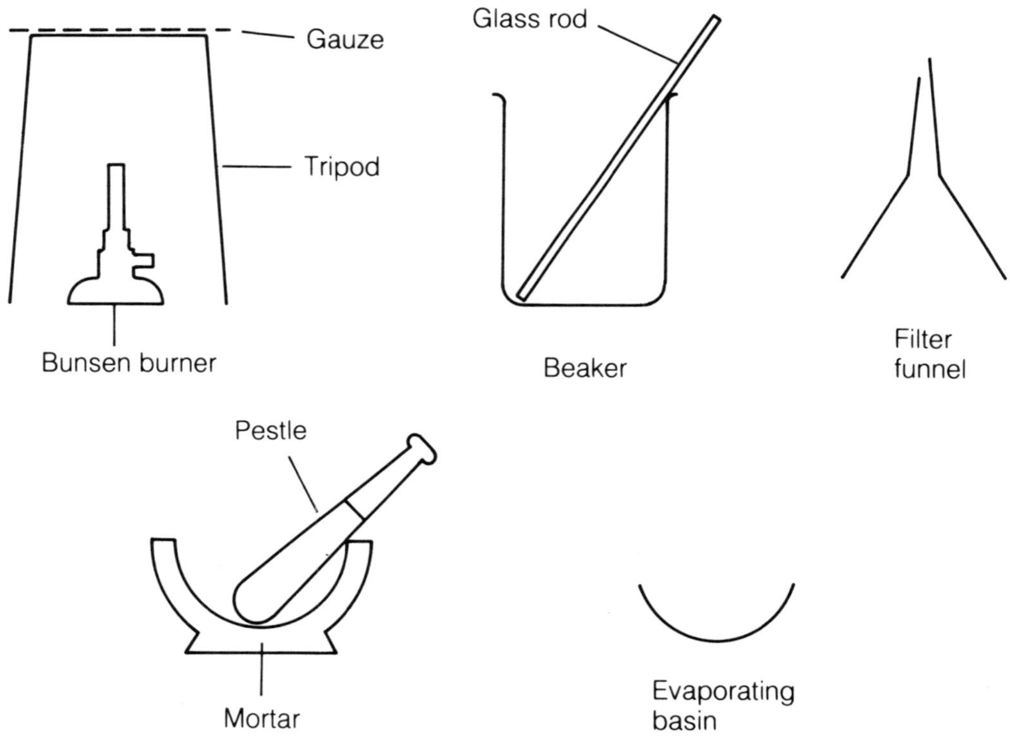

Fig. 2.1 Apparatus available

1 To help you to work out a suitable method of extraction, answer these questions. If you do not know the answers, find out by experiment.

Which dissolves faster, a lump of rock salt or powdered rock salt?
How can you turn a lump into a powder?
Does salt dissolve faster in cold water or warm water?
Does it help salt to dissolve if you stir the water?
Once you have got the salt into solution, how will you separate the solution from insoluble rock and sand? Sketch the apparatus you would use.

How will you obtain solid salt from the solution? Sketch the apparatus you would use.

2 When you have worked out how you can obtain pure salt from rock salt, write a brief plan. Show it to your teacher, and, if your plan is approved, put it into action.

3 When you have finished the experiment, copy this passage, filling in the blanks.

A solid dissolves in a liquid to form a ————. The solid is called the ————, and the liquid is the ————. Salt can be separated from sand and rock because salt is ———— in water, while sand and rock are ———— in water. The residue in the filter paper is ————. The filtrate (the liquid which filters through the filter paper) is ————. When the filtrate is evaporated, ———— vapour and ———— crystals are formed.

Experiment 2.2
Separating sodium chloride, lead(II) chloride and sand

1 Put a spatula measure of sodium chloride into a boiling tube. Add 20 cm^3 of water. Warm. If all the sodium chloride dissolves, add some more. When you have a saturated solution of sodium chloride, hold the boiling tube under the cold tap to cool it. Does sodium chloride come out of solution?

2 Repeat step **1**, using lead(II) chloride. Does lead(II) chloride come out of the solution when you cool it?

3 In a similar way, check whether sand dissolves in cold water and in hot water.

4 Record your observations in **1**, **2** and **3**. Make a note of the difference between the behaviour of sodium chloride, lead(II) chloride and sand. On the basis of your observations, plan a method of separating sodium chloride and lead(II) chloride and sand in a mixture of all three. Describe the steps that you would take. List the apparatus that you would need. If your teacher approves of your plan, carry out your experiment.

Experiment 2.3
Separating black ink into a solid and a liquid

In this experiment, you trap and condense the vapour formed when ink is heated.

1 Set up the apparatus shown in Fig. 2.3. Add some pieces of broken pot to prevent bumping (violent boiling). The side-arm boiling tube must have a close-fitting cork to hold the thermometer and a long delivery tube. This carries vapour out of the boiling tube into a cooled test-tube.

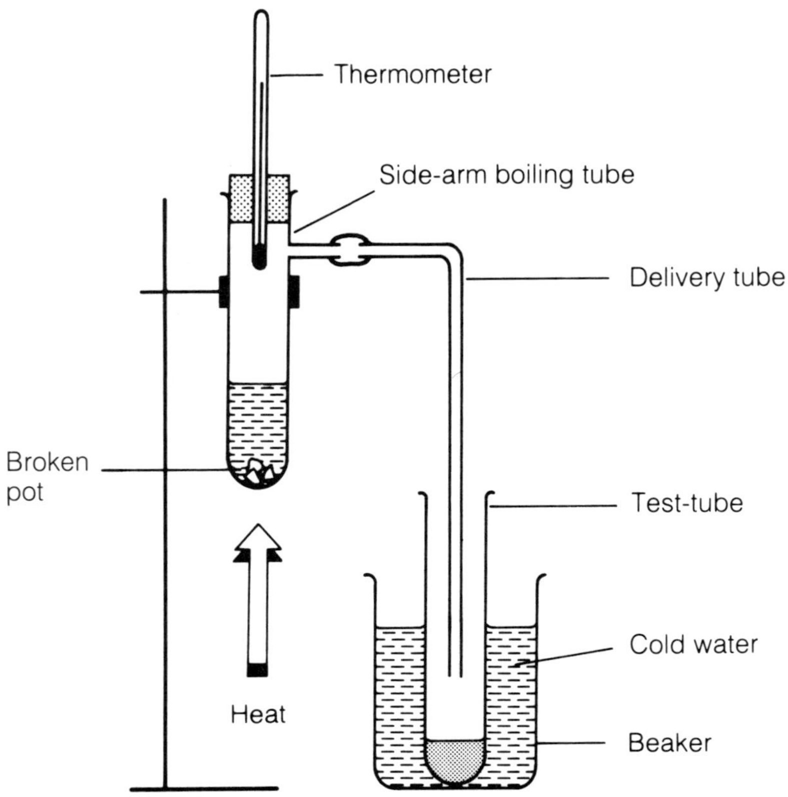

Fig. 2.3 Heating black ink

2 Heat the boiling tube, using a blue Bunsen flame.

3 Observe the distillate which collects in the test-tube.
Does the distillate look like ink?
What is its boiling point?
What do you think the distillate is?
Was the method of condensing the distillate efficient?
What improvements could be made to the method?

Experiment 2.4
Separating ethanol and water by distillation

You will have found that the method shown in Fig. 2.3 for condensing a vapour was inefficient. The present experiment uses a Liebig condenser. Distillation with a Liebig condenser can be used to separate two volatile liquids.

1 Set up the apparatus shown in Fig. 2.4. The distillation flask has a round bottom which assists even heating of the liquid inside it. Pour in some of the ethanol–water mixture. Add some pieces of broken pot to promote smooth boiling. Clamp the flask over a tripod and gauze. Insert a cork fitted with a thermometer so that the thermometer bulb is just below the side-arm of the flask. Bring the Liebig condenser into position carefully, and clamp it at the right height. Connect one piece of rubber tubing to the tap and put the other in a sink so that water flows in at the bottom of the condenser and out at the top. Have a beaker ready to receive any distillate.

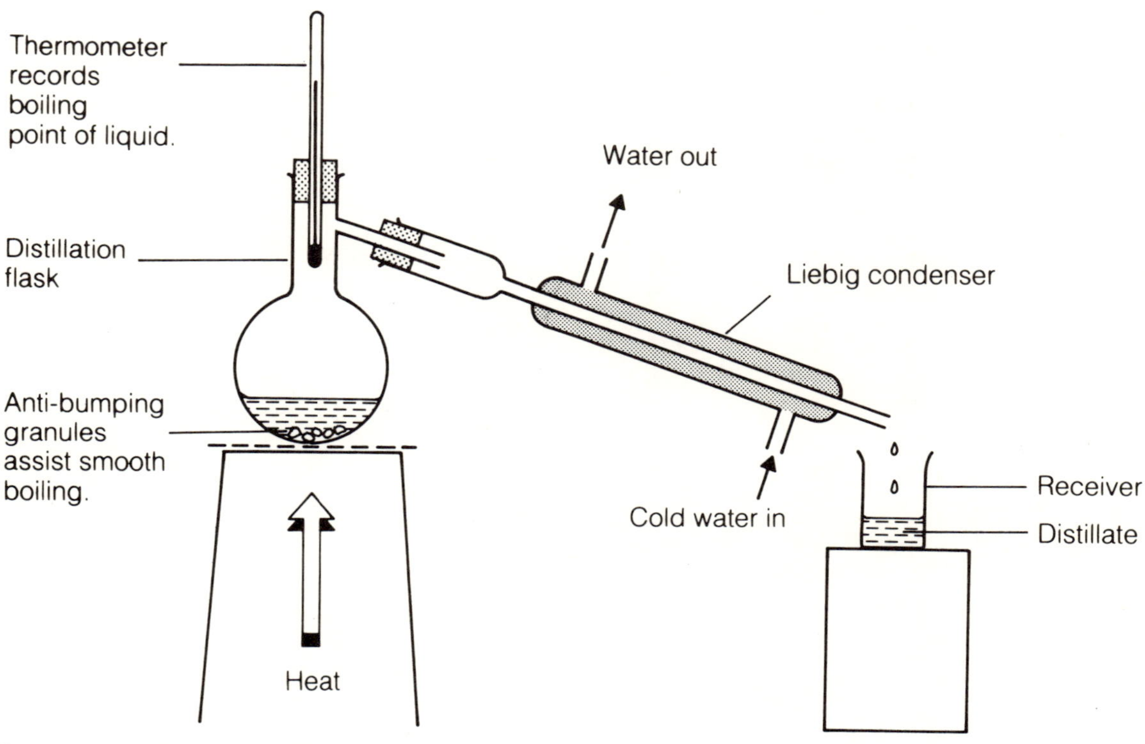

Fig. 2.4 Distillation apparatus

2 Heat the distillation flask. The temperature registered by the thermometer will rise until it reaches 78 °C. This is the boiling point of ethanol. The temperature remains steady while all the ethanol distils over.

3 When the temperature starts to rise above 78 °C, change the receiver.

4 When the temperature reaches 100 °C, water starts to distil over. Change the receiver again to collect water.

5 Do the following tests on each distillate.

a Smell it.

b Take a drop on the end of your finger, and taste it.

c Transfer 2 cm^3 with a teat pipette on to a watch glass. Test it with a lighted splint.

6 Copy the following passage, filling in the missing words or phrases.

The separation of a mixture of liquids into its components is called fractional distillation. In the distillation flask, liquid ————; in the condenser, vapour ————. Each liquid in the mixture distils at its ———— ————. A Liebig condenser is a piece of apparatus for ————————. It consists of two parts, an inner ———— whose purpose is —————— and an outer ———— whose purpose is ————. The reason why the temperature remains constant while a pure liquid is distilling is ————————.

Experiment 2.5
Is the pigment in black ink a pure substance?

In Experiment 2.4, you separated black ink into a solute and a solvent. You can now look at the solute, the pigment, to find out whether it is a pure substance or a mixture of pigments. You can use the method of *chromatography*. This experiment uses *ascending paper chromatography*.

1 Take a strip of chromatography paper 12 cm × 4 cm. Rule a pencil line 2 cm from the short side. Put a spot of black ink on the pencil line (see Fig. 2.5a).

2 With the spot at the bottom end, turn over 2 cm of paper at the top end. Staple this end round a glass rod (see Fig. 2.5b).

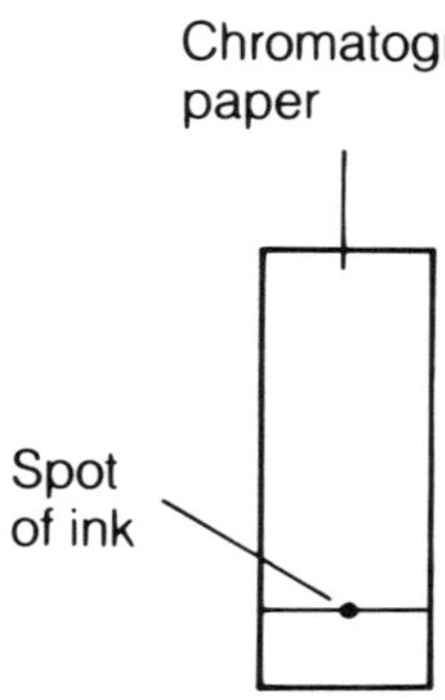

Fig. 2.5(a) Placing a spot of black ink on chromatography paper

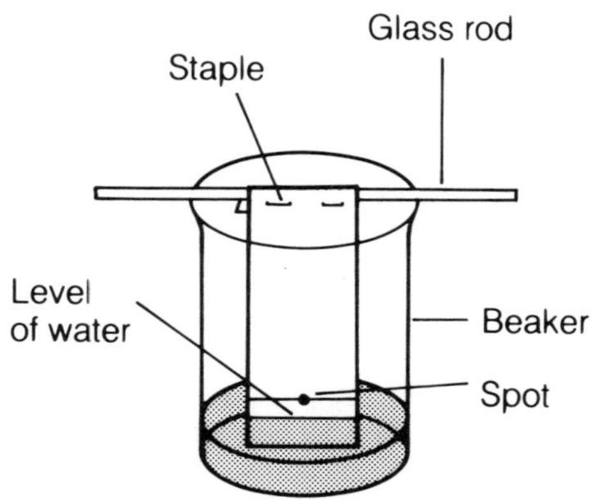

Fig. 2.5(b) Ascending paper chromatography

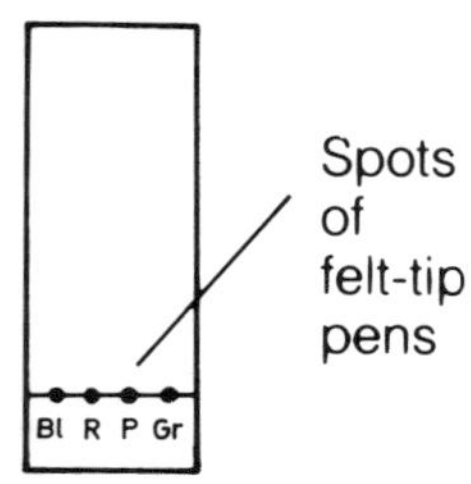

Fig. 2.5(c) Placing spots of different colours on chromatography paper

3 Pour water into a 250 cm^3 beaker to a depth of 1 cm. Position the glass rod across the top of the beaker. The ink spots will be 1 cm above the level of the water. This is important: water must travel up the paper to meet the ink spot; if the spot is below water level, the ink will just wash out. It is also important to make sure that the paper does not touch the side of the beaker.

4 When the water level has risen nearly to the level of the staple, remove the piece of chromatography paper and put it in an oven at 50 °C to dry.

5 You can now say whether the pigment in black ink is a pure substance or a mixture of dyes.

Extension 1 *Felt-tip pens*

You may like to repeat the experiment with inks of different colours. Use felt-tip pens to put spots of inks of different colours on a strip of chromatography paper. Write the colour of each ink in pencil underneath the ink spot (Fig. 2.5c).

Extension 2 *The indicator, screened methyl orange*

An indicator is a substance which has different colours in acidic, alkaline and neutral solutions. Screened methyl orange is an indicator. Find out whether screened methyl orange is a single dye or a mixture of dyes. You can do this by following the instructions for black ink.

Extension 3 *Smarties®*

Smarties® come in a number of colours. Plan an experiment to find out whether the dyes used to colour Smarties® are pure substances or mixtures. Say how you would remove samples of the dyes from the sweets and how you would attempt to separate them into components.

Carry out the experiment you have planned.

Experiment 2.6
Using chromatography to identify a dye

You will have found, if you have done Experiment 2.5, Extension 2, that the indicator, screened methyl orange, is a mixture of a yellow dye and a blue dye. In this experiment, you are told that the blue dye is bromophenol blue or methylene blue or xylene cyanol. You have to find out which of the three it is.

1 Make four glass jets. These are made by drawing out a piece of glass tubing in a Bunsen flame. **Wear safety glasses**. Hold a length of glass tubing (about 18 cm) between your forefingers and thumbs (see Fig. 2.6a). Hold the middle of the glass tube in a Bunsen flame. Rotate it until it looks yellow and feels soft. Take it out of the flame, and pull (see Fig. 2.6b). Break the thin drawn-out tubing in the middle so that you have two glass jets (see Fig. 2.6c). Use a second length of glass tubing to make two more glass jets.

(a) Glass tubing is rotated in the flame . . .

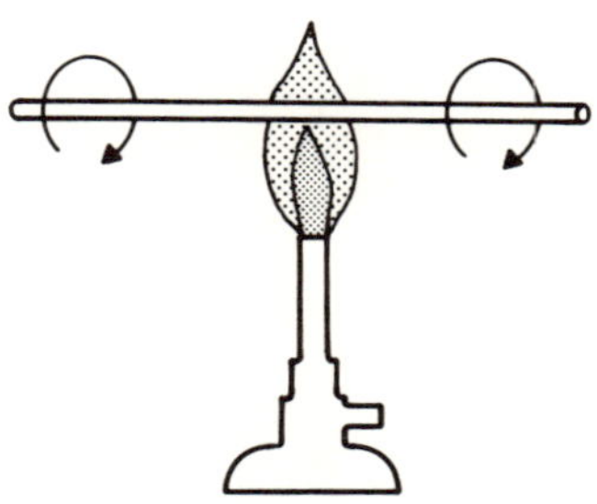

Fig. 2.6(a) Heating glass tubing

(b) then taken out of the flame and pulled . . .

Fig. 2.6(b) Drawing out glass tubing

(c) . . . and finally snapped into two pieces.

Fig. 2.6(c) Two glass jets

2 Take a strip of chromatography paper 12 cm × 12 cm. Draw a pencil line 2 cm from one end. Mark four points in pencil at intervals of 2 cm. Put them at 3 cm, 5 cm, 7 cm and 9 cm from the edge of the paper (see Fig. 2.6d). Underneath, write in pencil the names screened methyl orange, bromophenol blue, methylene blue and xylene cyanol.

3 Using glass jets, place spots of the four dyes on the marked points. Use a different jet for each dye. Let the spots dry. Then apply a second spot of each dye in the same place.

4 Curl the paper round and staple the edges together (see Fig. 2.6e).

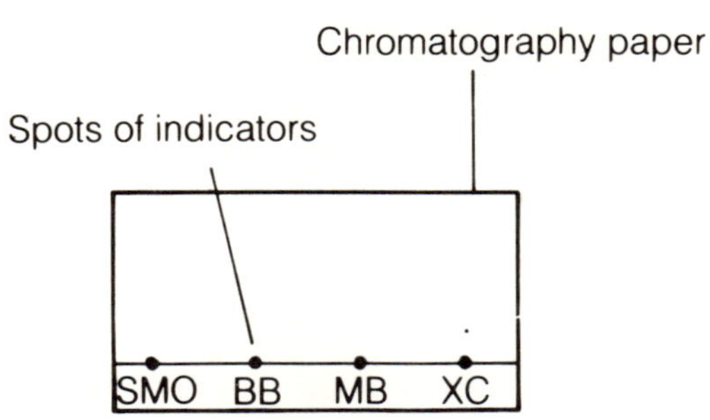

Fig. 2.6(d) Placing the spots

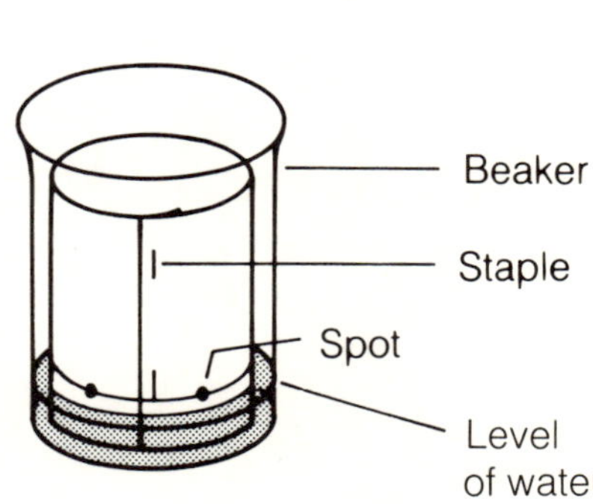

Fig. 2.6(e) Making the chromatogram

5 Take a 250 cm^3 beaker. Pour water into the beaker to a depth of 1 cm. Stand the cylinder of paper in the beaker. It is important that the spots are higher than the level of water in the beaker (see Fig. 2.6e).

6 When the water has risen to near the top of the paper, take the paper out of the beaker. Dry it in an oven at 50 °C.

7 What can you deduce about the blue dye in screened methyl orange?

Extension

You still do not know what the yellow dye is. Think up an experiment you could do to identify the yellow dye. Say what apparatus and chemicals you would need. Draw a sketch of the experiment you would do. If you can obtain the things you need, carry out your experiment.

Experiment 2.7
Chromatography on petals and leaves

Water is not the only solvent used in chromatography. The pigments in flower petals and in green leaves do not dissolve in water. This experiment on petals uses propanone as solvent. Antirrhinum, dahlia, petunia and rose petals and many other red, blue and purple petals give interesting results.

1 Take some flower petals. Put them into a mortar with 2–3 cm^3 of propanone. Grind them with a pestle until the pigment has been extracted by the solvent.

2 Use a teat pipette to put a spot of the extract on to a strip of chromatography paper 10 cm × 2 cm at a distance of 2 cm from the short end. When the drop has dried, add another drop in exactly the same place. Continue until you have built up a strongly coloured spot.

3 A good developing solvent is a mixture of propanone (1 volume) and water (3 volumes). Put 5 cm^3 of this mixture into a boiling tube. Hold it in your hand to vaporise some of the propanone and drive out air. Stopper the tube with a rubber bung.

4 Take the piece of chromatography paper. Fasten it with a drawing pin to the rubber bung (see Fig. 2.7). Hang the chromatography paper inside the boiling tube so that the spot of pigment is above the surface of the solvent and the bottom edge of the paper is just below the surface.

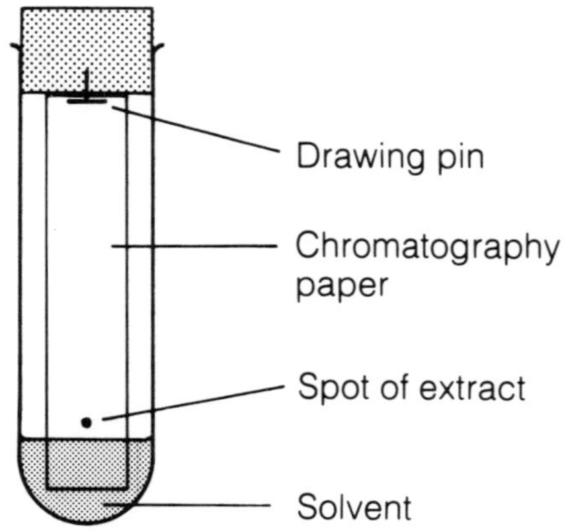

Fig. 2.7 Chromatography with a solvent other than water

5 Wait until the solvent has risen to the top of the paper. Take the paper out, and hang it up to dry.

6 Is the pigment in the petals a pure substance or a mixture?

7 Repeat the experiment, using a mixture of propanone (1 cm^3) and water (3 cm^3) and hydrochloric acid (3 drops) to develop the chromatogram. What effect does the hydrochloric acid have on the separation? What other effect does it have? Try to explain what you see.

Extension

The pigment in green leaves can be investigated in a similar way. Propanone and ethanol are both good solvents to use for extracting the pigment. Nettles and privet leaves are good sources of pigment. If you dry them before use, you can make a more concentrated extract, and obtain better results.

Experiment 2.8
Column chromatography on food colourings

Chromatographic separations can be made by running solutions of mixtures through columns of adsorbent material. Aluminium oxide, silicon(IV) oxide and cellulose are all used as adsorbents. The present experiment uses cotton wool. Both cotton wool and paper consist of cellulose. The column is a test-tube with a hole in the end. (An alternative is a length of glass tubing drawn out into a jet at one end as described in Experiment 2.6.) You will get interesting results with some food colourings, especially black and blue dyes, and with screened methyl orange.

1 Take a test-tube with a hole in the end (Teacher: see note about Experiment 2.8 in Appendix). Support it in a clamp as shown in Fig. 2.8a.

2 Shred some cotton wool, put it into a beaker, and mix it thoroughly with water. Pour some of the mixture of cotton wool and water into the test-tube. Collect the water which runs out through the hole in the test-tube. Using a glass rod, pack the cotton wool down inside the test-tube. Build up a column 8 cm high (see Fig. 2.8a).

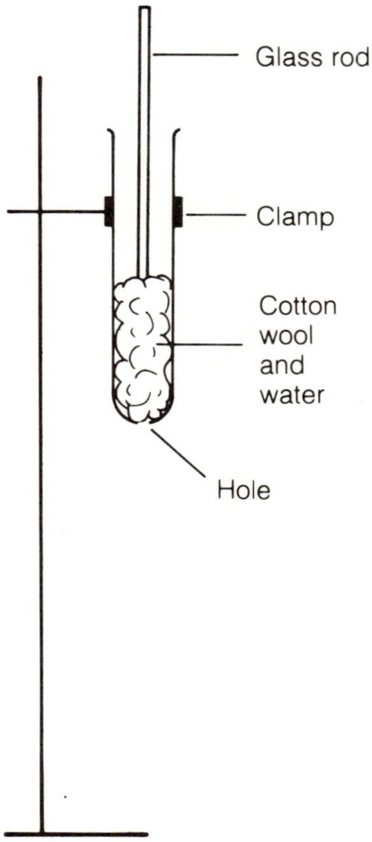

Fig. 2.8(a) Preparing the column

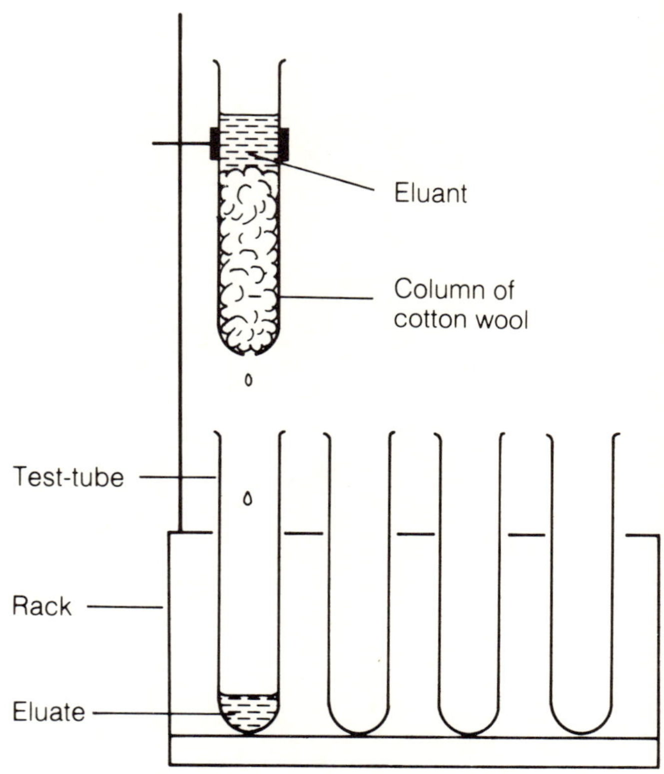

Fig. 2.8(b) Collecting the eluate

3 Wait until the surplus water has run out of the tube. Use a teat pipette to add 10 drops of food colouring. Wait until it has soaked into the cotton wool.

4 Add water to fill the tube. As water drains out of the bottom, keep the tube topped up. Collect the water which runs out of the column in a beaker. When the eluate (the water leaving the column) becomes coloured, start to collect it in a rack of test-tubes. Collect 5 cm^3 of eluate in each test-tube. Continue until all the coloured eluate has been collected (see Fig. 2.8b).

5 How many portions of coloured eluate have you collected?
How many different colours can you see in the portions?
How could you obtain solid samples of the dyes present in the food colouring?

Extension

You may like to try some different solvents, as follows.

- **a** For Smarties®, a mixture of butan-1-ol: ethanol: ammonia (3:1:1 by volume) gives good results. Dark brown Smarties® will give an exceptionally interesting result.
- **b** For food colourings, e.g. Boots' food colours, a mixture of butan-1-ol: ethanoic acid: water (11:4:5 by volume) gives good separations.
- **c** A mixture of propanol and ammonia (1:1 by volume) also gives good results with food colours.

Experiment 2.9
Plotting a cooling curve for a molten solid

1 Set up the apparatus shown in Fig. 2.9a. Octadecanoic acid (stearic acid) is a solid with a suitable melting point.

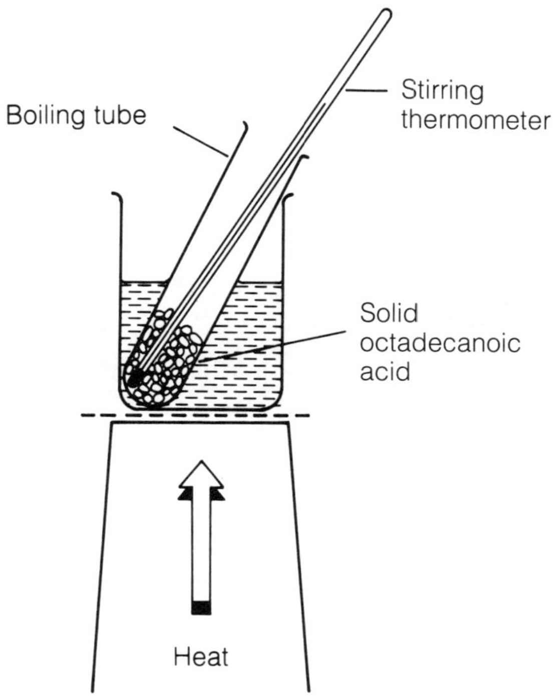

Fig. 2.9(a) Melting point apparatus

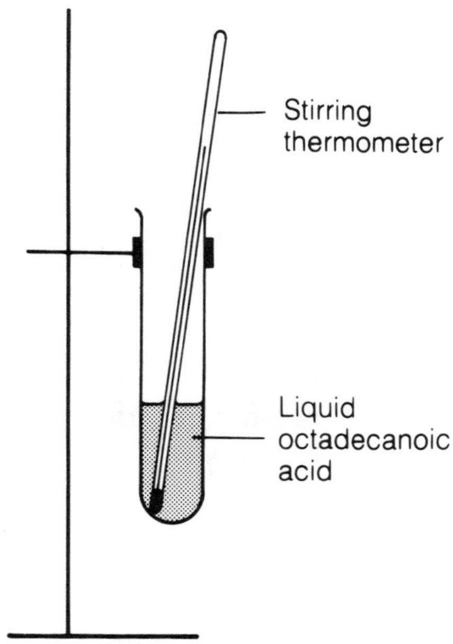

Fig. 2.9(b) Cooling

2 Heat the water in the beaker. When the solid octadecanoic acid in the boiling tube melts, turn off the Bunsen.

3 Remove the boiling tube from the hot water. Clamp it vertically (see Fig. 2.9b). Record the temperature of the liquid.

4 Stir the liquid gently as it cools. Watch the thermometer. Record the temperature at intervals of ½ minute until 2 minutes after all the liquid has solidified.

5 Tabulate your results as shown below.

Table: Cooling of a liquid

Time since you stopped heating	*Temperature of liquid in °C*

6 You need to remove the thermometer from the solid acid. Put the boiling tube into the hot water again until the solid melts and then take out the thermometer.

7 Plot your results as shown in Fig. 2.9c.

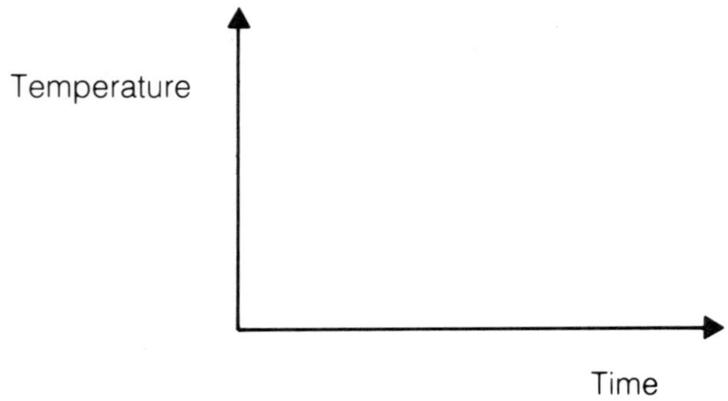

Fig. 2.9(c) How to plot a cooling curve

8 Explain why the graph has the shape you have obtained. Mark on it the melting point of the solid.

9 What difference does it make if you do not stir during cooling? To find out, repeat step **3** without stirring. Tabulate and plot your results as in steps **4** and **5**.

10 Repeat the experiment with candle wax.
Say whether the graph that you obtain has the same shape as the one that you obtained for octadecanoic acid.
If the candle wax graph is a different shape, explain why it is different.

11 Urea has a melting point of 133 °C. What alteration to this method would you have to make in order to check the melting point of a sample of urea?
If the sample melted at 130 °C, what would you deduce?

TOPIC 3 Elements and compounds

Experiment 3.1
What happens when iodine meets hot iron?

Iron is a metallic element, and iodine is a non-metallic element.

1 Take a Pyrex test-tube. Put 5 flakes of iodine in the bottom of it. Push a wad of iron wool half-way down the test-tube.

2 Clamp the test-tube at its open end (see Fig. 3.1).

3 Heat the iron wool strongly. From time to time, move the Bunsen flame to heat the iodine and drive iodine vapour over the heated iron.

4 When all the iodine has reacted, and you can no longer see the colour of iodine vapour in the test-tube, turn out the Bunsen and allow the test-tube to cool.

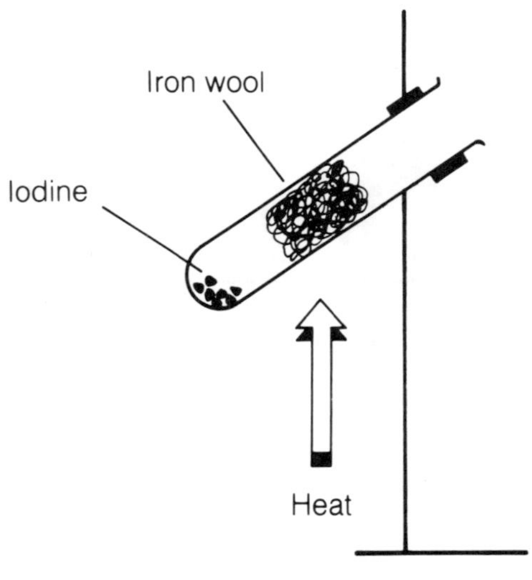

Fig. 3.1 Heating iron and iodine

5 Describe the appearance of the iron wool before it was heated.
Describe **a** solid iodine, **b** iodine vapour.
Describe the appearance of the iron wool after the experiment.

Important parts of science are *observation* and *deduction*. You have to observe (see) changes and to deduce (work out) the reason. From your observations, what do you deduce has happened to iron and iodine?

Experiment 3.2
Mixing tin and iodine

In this experiment, the solvent dichloromethane is used.
1,1,1-Trichloroethane can be used instead, but it takes longer to evaporate.

1 Take 10 cm^3 of dichloromethane in a dry boiling tube. Add two flakes of iodine and a piece of tin foil (1 cm × 2 cm).

2 Stand the boiling tube in a beaker of warm water at 30 °C for 1 minute (see Fig. 3.2a). The solvent boils at 40 °C.

3 Remove the boiling tube, and let it stand for 3 minutes.

4 Decant the liquid on to a crystallising dish (see Fig. 3.2b).

5 Warm the crystallising dish over hot water (see Fig. 3.2c). **Do not use a flame because the solvent is flammable**. When the solution starts to crystallise, remove it from the heat, and leave it to stand.

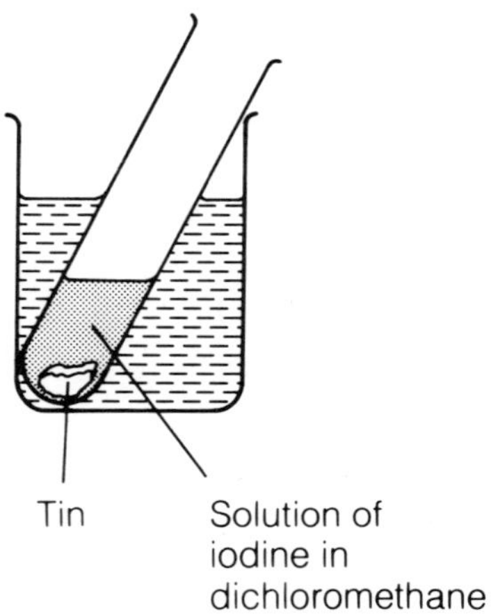

Fig. 3.2(a) Heating the mixture to 30 °C

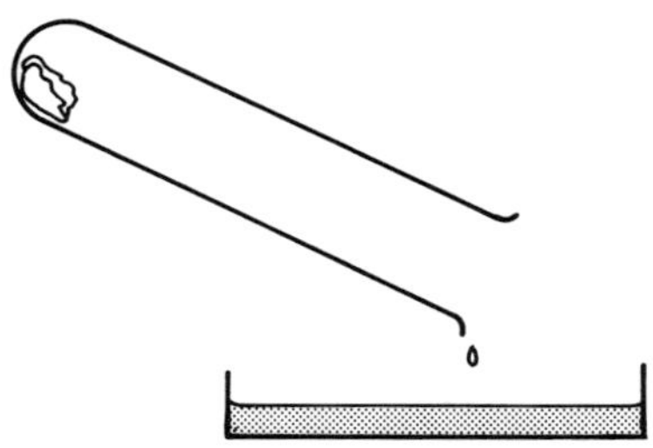

Fig. 3.2(b) Decanting the solution

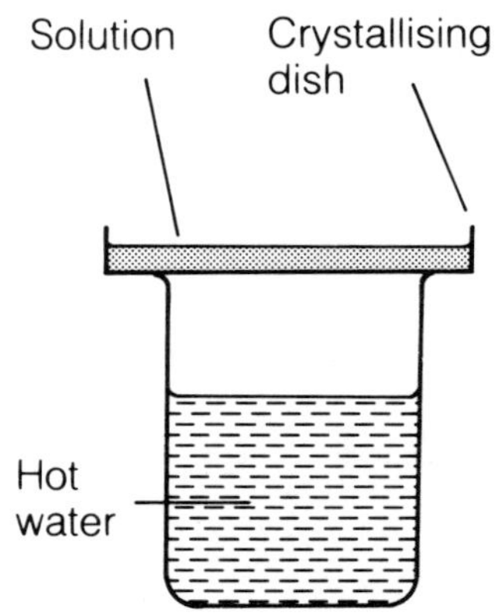

Fig. 3.2(c) Warming the crystallising dish over hot water

6 Describe the appearance of the crystals that you obtain.
Describe the appearance of tin and iodine.
What do you think the crystals are?
Write a word equation for the reaction that has happened.

Experiment 3.3
What happens when sulphur meets hot copper?

Sulphur is a non-metallic element; copper is a metallic element.

1 Take a Pyrex test-tube. Tip one spatula measure of sulphur into it.

2 Cut a piece of copper foil 1 cm × 4 cm in size. Weigh it, and write down its mass.

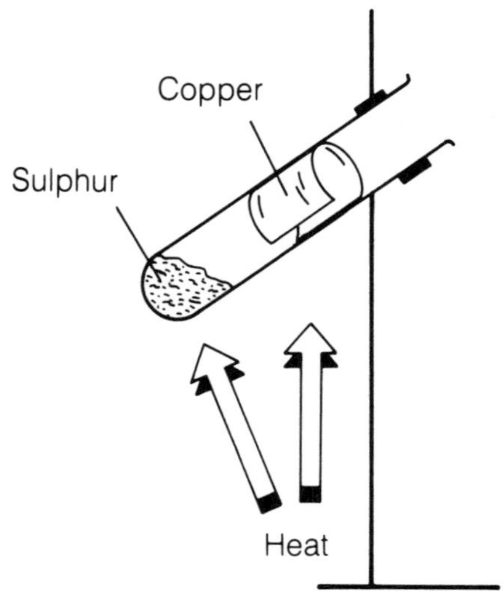

Fig. 3.3 Heating copper with sulphur

3 Push the strip of copper half-way down the test-tube. Only the sides of the strip should touch the tube. Hold the test-tube in a clamp, as shown in Fig. 3.3.

4 Use the Bunsen flame to heat both the copper and the sulphur. The idea is to drive sulphur vapour over heated copper. When the copper starts to glow, move the flame away from the copper and position it under the sulphur. This will keep a stream of sulphur vapour flowing over the copper.

5 When nothing further seems to be happening, let the remaining sulphur cool. Then heat the copper to vaporise any sulphur that has solidified on it.

6 Let the copper cool. Carefully lift the strip of copper out of the test-tube. Reweigh it, and write down its mass.

7 Describe the strip of copper that you started with.
What changes do you find in its appearance and behaviour?
Has the mass increased or decreased or stayed the same?
Is it still copper? If not, what is it?

Write a summary of the changes that have occurred. If you think that a chemical change has occurred, write a word equation for it.

Experiment 3.4
Heating a mixture of iron and sulphur

1 Weigh 7 g of iron. Weigh 4 g of powdered sulphur. Mix the two elements together. **Do not grind them together**.

2 Tip half the mixture into a Pyrex test-tube. (Keep half for step **5**.) Clamp the tube at its open end as shown in Fig. 3.4a.

3 Heat the mixture at the closed end of the test-tube. When the mixture starts to glow red, take away the Bunsen. Make a note of what you see.

4 When the test-tube has cooled, scrape out its contents. If you cannot remove the contents from the tube, you will have to smash the tube. First

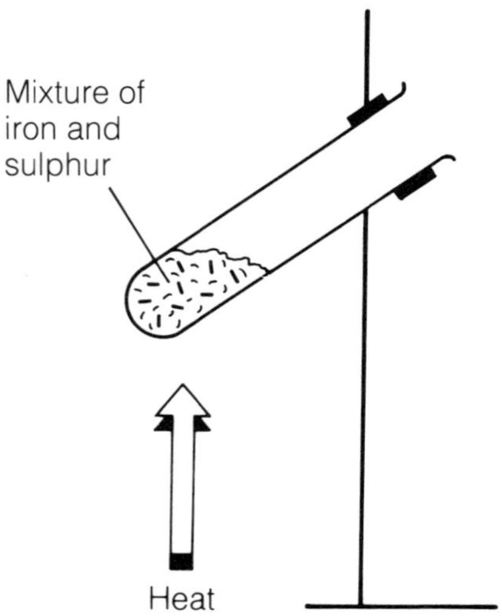

Fig. 3.4(a) Heating a mixture of iron and sulphur

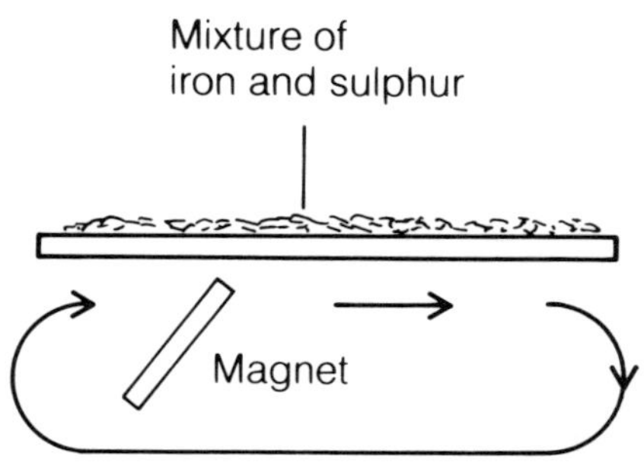

Fig. 3.4(b) Testing the mixture with a magnet

wrap it in a piece of paper towel to collect the bits of broken glass. **Wear your safety glasses**.

5 Do the following tests on (i) the unheated mixture and (ii) the contents of the test-tube.

- **a** Observe the colour and texture.
- **b** Put a little of the substance into a test-tube. Half fill the test-tube with water. Shake it.
- **c** Sprinkle some of the substance on to a piece of paper. Bring a magnet under the paper, and move it towards one end of the paper, as shown in Fig. 3.5b. Repeat.

  Tabulate your results as shown in the table.

	Mixture of iron and sulphur	*Mixture after heating*
Appearance		
Shaking with water		
Action of magnet		

6 What did the red glow that spread through the test-tube, even though you had stopped heating, tell you?

7 Is the substance that you took out of the test-tube still a mixture of sulphur and iron? If not, what do you think it is? Write a summary of the change that has occurred.

TOPIC 4 Radioactivity

Exercise 4.1
Plotting radioactivity against time

The radioactivity of a radioactive isotope can be measured as counts per minute (c.p.m.) registered by a Geiger counter. The radioactivity in c.p.m. decreases with time. Plot the values of c.p.m. shown in the table against time. Use your graph to find the half-life of the radioisotope.

Table: Decrease in radioactivity with time

Time in minutes	*Radioactivity in c.p.m.*
2.5	107
5.0	89
10.0	65
15.0	48
20.0	34
25.0	25
30.0	16
35.0	10
40.0	8
45.0	6
50.0	5

TOPIC 5 Electrochemistry

Experiment 5.1
Finding out which substances conduct electricity

1 You are supplied with a 6 V battery or a transformer which gives direct current at 6 V, a bulb in a bulb-holder, a switch, electrical wires ending in crocodile clips and a collection of substances to test.

2 Assemble an electrical circuit which will enable you to find out whether a solid is an electrical conductor. Make a drawing of your test circuit.

3 Test the metals aluminium, copper, iron, lead, tin and zinc.

4 Test lumps of the non-metallic elements sulphur, charcoal and graphite. The last two are forms of carbon.

5 Find out whether a crystal of copper(II) sulphate will conduct electricity.

6 Next test the crystalline substances copper(II) chloride, potassium iodide, lead bromide and sodium chloride. These substances are composed of small crystals, and you will need to test them as shown in Fig. 5.1a.

7 Now test some liquids. Assemble the apparatus shown in Fig. 5.1b. Grip two carbon electrodes with crocodile clips. Slot the carbon rods through holes in a piece of cardboard which rests on top of a beaker. This will prevent the rods from touching. Into the beaker, pour some distilled water. Connect the carbon rods to your test circuit, and find out whether water is an electrical conductor.

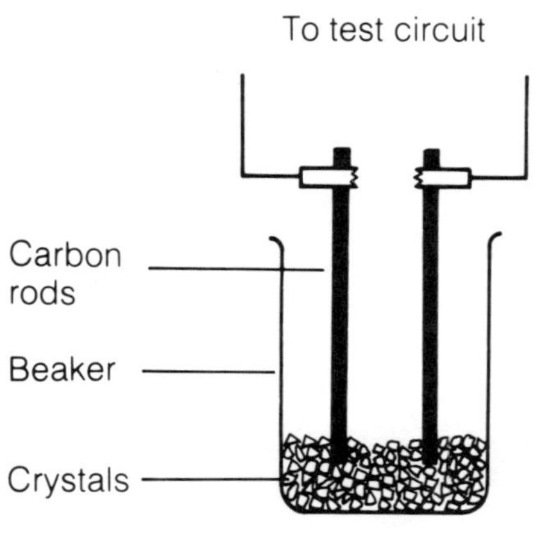

Fig. 5.1(a) Apparatus for testing crystals

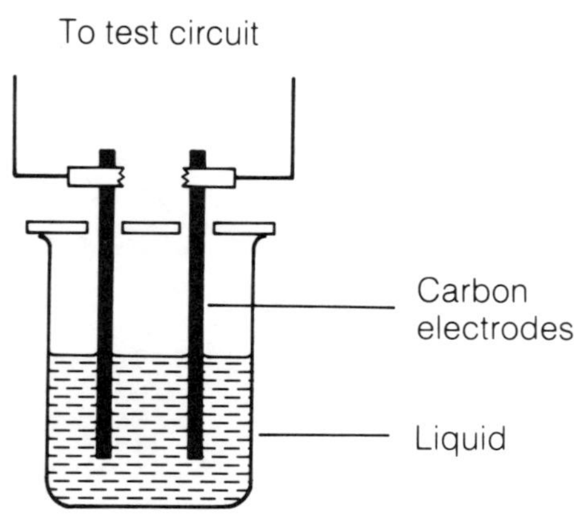

Fig. 5.1(b) Apparatus for testing liquids

8 Repeat with the liquids ethanol, glycerol and propanone. Repeat with solutions of sulphuric acid, ethanoic acid, copper(II) sulphate, copper(II) chloride, sodium chloride, sugar and ammonia.

9 Tabulate your observations (Tables A and B).

Table A Conduction of electricity by solids

Solids which conduct electricity	*Solids which do not conduct*

Table B Conduction of electricity by liquids

Liquids which conduct electricity	*Liquids which do not conduct*

Experiment 5.2
What happens when solutions conduct electricity?

1 Set up the apparatus shown in Fig. 5.2. If you are using one of the small commercial glass electrolysis cells, place ignition tubes over the electrodes. The apparatus shown here uses test-tubes because the electrolysis cell is larger. The cell is made from a plastic juice bottle or a plastic chemical reagent bottle. A two-hole rubber bung holds the carbon electrodes. Fill the apparatus with copper(II) chloride solution.

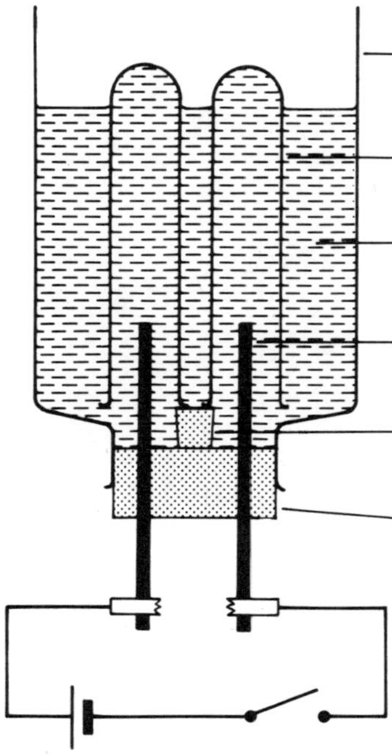

Fig. 5.2 Apparatus for electrolysis

2 When the current is flowing, and the bulb lights, remove the bulb and holder from the circuit to allow more current to flow.

3 If you see a gas being evolved, place a test-tube full of solution over the electrode, and collect the gas.

4 Make a note of the changes which occur at the positive electrode (anode) and the negative electrode (cathode).

5 Repeat with solutions of copper(II) sulphate, sodium chloride, potassium iodide and dilute sulphuric acid.

6 Make a note of all the changes you observe, and enter them into a table.

Table of results

Solution	*Seen at positive electrode*	*Seen at negative electrode*

Experiment 5.3
Why does electricity make some substances move?

1 Set up the apparatus shown in Fig. 5.3. Moisten the strip of filter paper with sodium chloride solution.

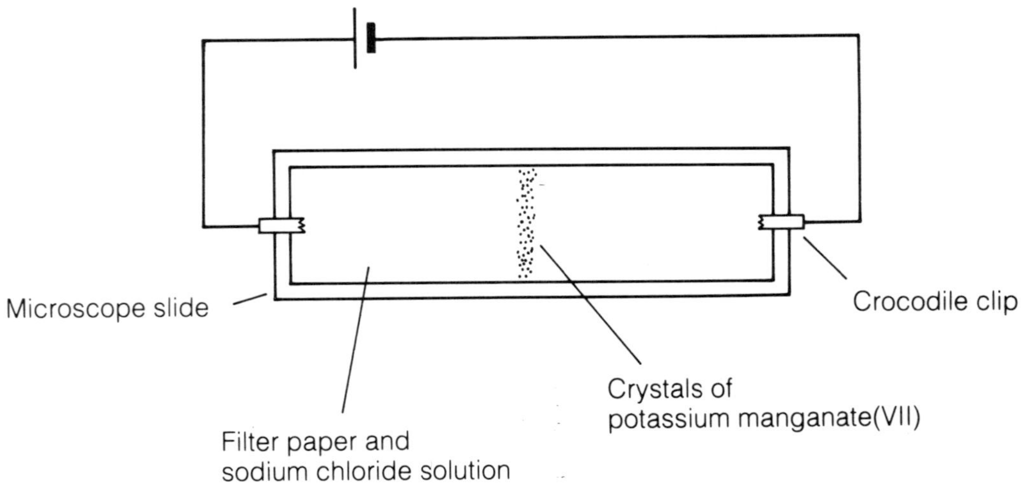

Fig. 5.3 Looking for movement

2 Carefully make a thin line of potassium manganate(VII) crystals in the centre of the damp filter paper.

3 Switch on the 6 V battery. Let the current pass for 15 minutes.

4 What do you observe? Is the purple colour still in the line where you applied the crystals? Has it moved towards either the positive electrode or the negative electrode? Can you explain what happens?

5 Repeat the experiment, using a thin line of copper(II) sulphate crystals. At the end of 15 minutes, switch off. Take the stopper out of a bottle of dilute ammonia solution. Let the ammonia gas which comes from the open bottle come into contact with the filter paper strip. It will intensify the blue colour of the copper compound. What do you observe? Has the blue colour moved from the line where you applied the crystals? How does your result differ from what you saw in **4**? Can you explain the difference?

6 Potassium chromate(VI) and potassium dichromate(VI) can be used in other experiments.

7 Potassium manganate(VII), $KMnO_4$, contains the ions K^+ and MnO_4^-. What happens to the ions during the experiment?
Copper(II) sulphate, $CuSO_4$, contains the ions Cu^{2+} and SO_4^{2-}. What happens to the ions during the experiment?
What ions are present in (a) potassium chromate(VI), K_2CrO_4, and (b) potassium dichromate(VI), $K_2Cr_2O_7$? What colour are they? What happens to the ions during the experiment?

Experiment 5.4
Combining zinc and iodine and passing a direct current through the product

1 Take a 100 cm^3 beaker. Tip into it one spatula measure of zinc dust (1.5–2.0 g). Add 50 cm^3 of water.

2 Take one spatula measure of iodine crystals. Grind them in a pestle with a mortar. **Do not grind zinc and iodine together**.

3 Tip the iodine into the water in the beaker. Stir. Watch the colour change as iodine and zinc react. You will see a residue of zinc.

4 When the solution has become colourless, filter it.

5 Pour the solution into a 100 cm^3 beaker. Pass a current through it in the apparatus shown in Fig. 5.4. Note the change at the negative electrode (cathode). Note what happens at the positive electrode (anode).

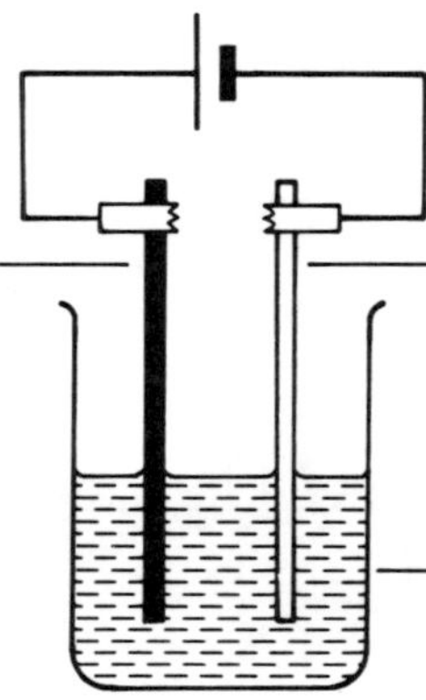

Fig. 5.4 Passing a current through a solution

6 Now explain what happened in step **3** and in step **5**. Copy and complete this passage.

When zinc and iodine were mixed in water, we saw ——————. Zinc and iodine are elements; zinc is a ———— element, and iodine is a ———— element. The chemical reaction that happened was

zinc + iodine → ————

When we passed a current through the solution, we saw certain changes. At the negative electrode (cathode), we saw ————. At the positive electrode (anode), we saw ————. This proves that the solution has been ———— by the electric current. The chemical reaction that has occurred is ——————.

Experiment 5.5
Electrolysis of copper(II) sulphate solution with copper electrodes

1 Clean two strips of copper foil. Sandpaper them thoroughly.

2 Set up the circuit shown in Fig. 5.5. Allow current to flow for 20 minutes.

3 Switch off.
What has happened at the negative electrode (cathode)?
What change do you see in the solution?
What has happened at the positive electrode (anode)? This is less easy to spot. Rub the surface of the positive electrode with your finger. It has changed since you sandpapered it. Try to explain what has happened to it.

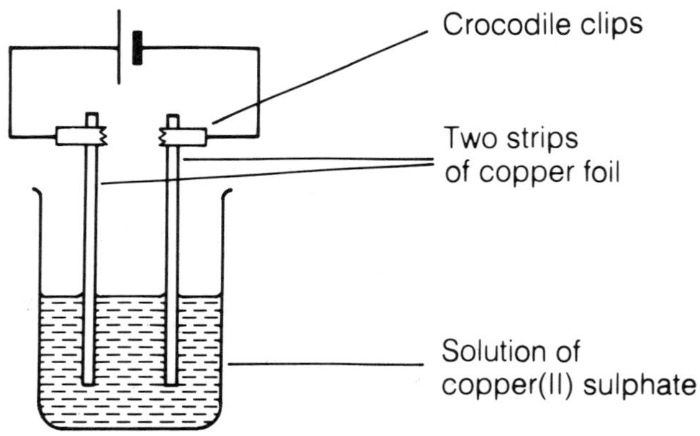

Fig. 5.5 Electrolysis of copper(II) sulphate

4 Formulate a theory to explain what you have observed at the electrodes and in the solution. How could you confirm your theory? Describe an experiment you could do. What measurement would you make of each electrode before and after the passage of the electric current? Write a detailed plan of the experiment you have in mind, and show it to your teacher. If he or she approves, carry out the experiment.

Experiment 5.6
A test circuit

Problem

You are given solutions of sodium sulphate, sodium hydrogencarbonate, ethanoic acid, propanol and glucose. You have to divide them into strong electrolytes, weak electrolytes and non-electrolytes. You have a battery and electrical connectors but no light bulbs.

Hint

During electrolysis, a gas is often given off.

Apparatus

You have beakers, cardboard, specimen bottles, corks, pieces of nichrome wire, pencil leads, a 9 V battery, electrical wire, crocodile clips, pieces of wood and a stapler.

1 Draw an apparatus in which you could test the solutions. Say what experiments you would need to do.

2 Construct the apparatus you have drawn, and do the tests you have described.

3 Divide the solutions into strong electrolytes, weak electrolytes and non-electrolytes.

Experiment 5.7
Copper plating (and zinc plating)

In this experiment, a layer of copper is deposited on an iron key or nail.

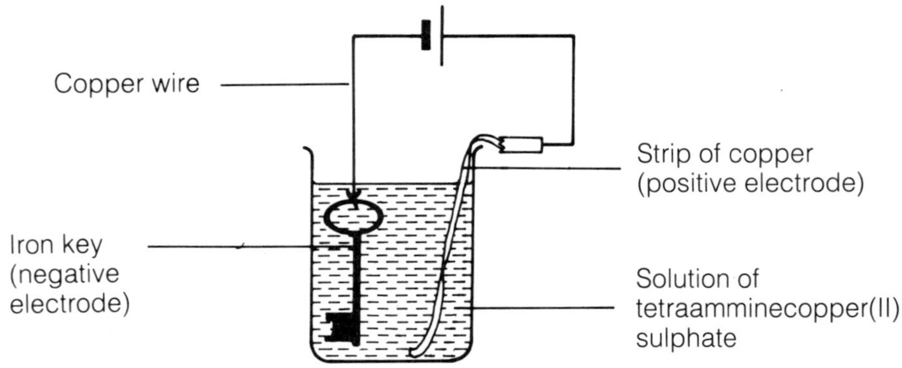

Fig. 5.7 Copper plating

1 First, clean the iron key or nail. Rub it with sandpaper or steel wool. Then place it in hot dilute hydrochloric acid solution for 2 minutes. Next, soak it in hot dilute sodium hydroxide for 2 minutes. Finally, rinse it with water.

2 Make a solution of tetraammine copper(II) sulphate. Take 1 g of copper(II) sulphate in 30 cm^3 of water in a 250 cm^3 beaker. Add dilute ammonia solution. A precipitate forms. Add more ammonia solution until the precipitate redissolves.

3 Attach a copper wire to the key or nail. Hang the key in the solution, as shown in Fig. 5.7. Attach the wire to the negative end of a 4.5 V battery.

4 Bend a strip of copper over the side of the beaker. Connect it to the positive end of the battery.

5 The electrical circuit is now complete. Let the reaction take place for 20 minutes.

6 Take out the key. Rinse it with distilled water.

7 If possible, dip the key into ethanol (or methylated spirit) before allowing it to dry. The reason for this is that the copper layer darkens if it is exposed to the air while damp.

Extension

How would you electroplate copper with zinc?

You are given a piece of copper foil, a zinc rod, zinc sulphate, a battery or a transformer, electrical wires and crocodile clips, and glassware. You are given the hint that, if you add a few drops of dilute sulphuric acid and some boric acid to the electrolyte, you will get better results.

1 Say what you would need to clean the copper foil.

2 Draw the electrolysis apparatus you would set up.

3 Make a note of any precautions you would take to obtain a good result.

4 Show your plan to your teacher, and, if he or she approves, carry it out.

Experiment 5.8
The lead–acid accumulator

1 Take two thin sheets of lead, 2 cm × 8 cm.

2 Soak the sheets of lead in sodium hydroxide solution for 10 minutes. This removes grease. Rinse them in water.

3 Grip the lead strips with the crocodile clips attached to a piece of wood. They must not touch (see Fig. 5.8).

4 Fill the beaker with dilute sulphuric acid.

5 Connect the crocodile clips to a 4.5 V battery. Pass a current for 3 minutes.

6 Disconnect the battery. Connect the leads to a 1.25 V bulb. See how long the bulb remains lit.

7 The lead and sulphuric acid are acting as an accumulator. When the bulb goes out, you can recharge the accumulator by repeating step **5**.

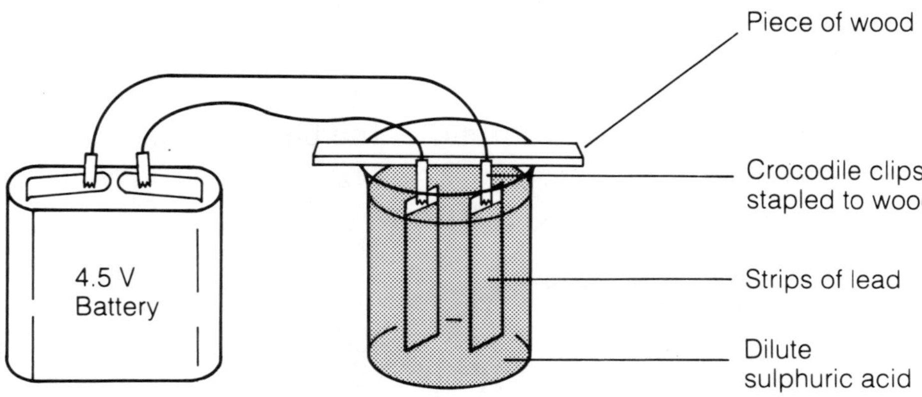

Fig. 5.8 A model lead–acid accumulator

8 The time for which the battery will light a bulb depends on a number of factors. Plan and carry out experiments to see how this time depends on each factor in the following list:

- **a** the time for which the accumulator is charged,
- **b** the distance between the pieces of lead,
- **c** the voltage used to charge the accumulator,
- **d** the concentration of sulphuric acid,
- **e** the volume of sulphuric acid,
- **f** any other factor that occurs to you.

To make sense of your results, you can change only one factor at a time. Bear this in mind when you are designing your experiments.

Experiment 5.9
Anodising and dyeing aluminium

When a piece of aluminium meets the air, it becomes coated with a surface layer of aluminium oxide. Heating in air makes the layer thicker. A much thicker layer of aluminium oxide can be made by *anodising* aluminium. This experiment is about a method of making the surface layer of aluminium oxide thicker. After it has been anodised (made the positive electrode, the anode, in a cell), the surface layer is able to absorb dyes.

1 Cut two pieces of aluminium foil, one 3 cm × 10 cm and the other 20 cm × 10 cm.

2 Soak them in a detergent to remove grease. Rinse with water.

3 Soak them in warm sodium hydroxide solution for 1 minute in a fume cupboard. This removes the existing oxide layer. Rinse with water.

4 Dip them into dilute nitric acid. Rinse with water.

5 Assemble the apparatus shown in Fig. 5.9. The larger piece of foil is bent into a cylindrical shape. The smaller piece hangs inside it. The two pieces must not touch. Attach crocodile clips to both pieces of foil.

6 Connect the cylinder to the negative terminal of the battery. Connect the strip of foil to the positive terminal of the battery. This makes the strip the *anode* of the cell.

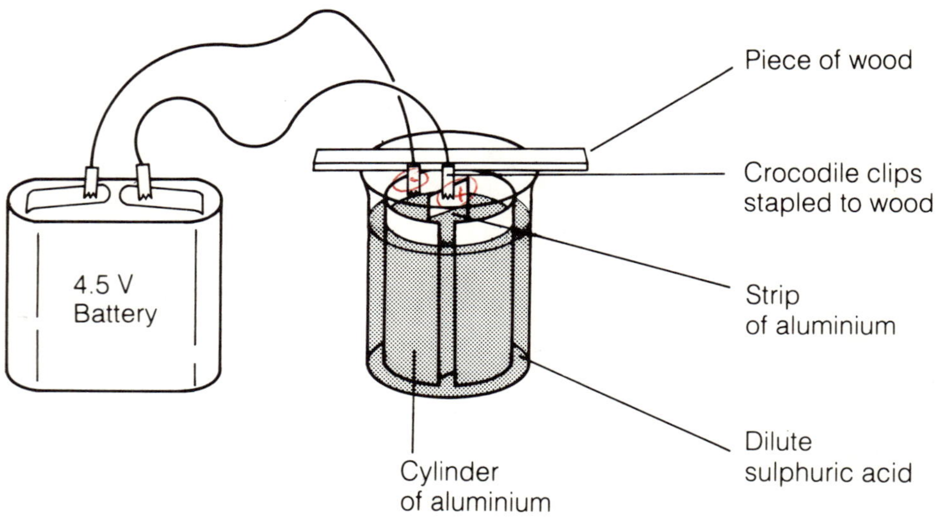

Fig. 5.9 Anodising aluminium

7 Pass a current for 20 minutes.

8 Make a solution of the dye Congo red. (Take 1 g of Congo red, 100 cm^3 of ethanol and 100 cm^3 of water in a beaker.) Heat the solution to 65 °C. Leave the anodised aluminium strip in the dye solution for 10 minutes.

9 Seal the dye. Immerse the dyed aluminium in boiling water for 10 minutes. This treatment makes the layer of aluminium oxide less porous.

10 You should do a control experiment to find out whether the cathode will take up dye. Repeat steps **8** and **9**, using the cathode (the negative electrode).

11 Repeat the anodisation and dyeing, but reverse steps **8** and **9**. Does anodised aluminium take up dye if it is sealed first?

12 You can see dyed anodised aluminium in use in door frames and window frames. Explain what advantages this material has over

a steel,

b aluminium,

c anodised aluminium which has not been dyed.

Experiment 5.10
Writing with electricity

1 Dissolve a spatula measure of sodium chloride in 2 cm^3 of water. Add 3 drops of phenolphthalein.

2 Lay a strip of filter paper on top of a glass slide as shown in Fig. 5.10. With a teat pipette, drop the solution on to the paper until the paper will hold no more.

3 Attach the positive end of a 6 V battery to a lead ending in a crocodile clip. Use the crocodile clip to grip one end of the paper.

4 Attach the negative end of the battery to a pencil 'lead'.

5 Holding the pencil 'lead' in your fingers, write lightly on the wet paper.

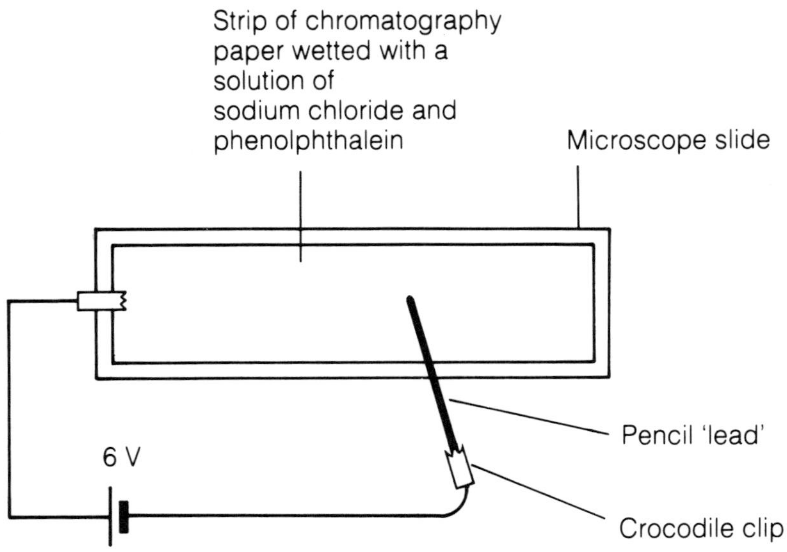

Fig. 5.10(a) Electric wiring

What colour is the writing? Explain what has happened to the sodium chloride solution to produce this colour.

6 Repeat the experiment. This time, add blue litmus solution in step **1**, instead of phenolphthalein. Attach the pencil 'lead' to the positive end of the battery. What colour is the writing this time? Explain what reactions have occurred to produce this colour.

7 Repeat the experiment, this time adding red litmus solution in step **1**. Attach the pencil 'lead' to the negative electrode. Describe what you see, and explain how you have produced this colour.
Note: Carbon electrodes can replace pencil 'leads'.

Extension 1

Other substances which you can use to get coloured writing are
a bromocresol green (with the 'lead' attached to the positive terminal),
b methyl orange (with the 'lead' attached to the negative terminal),
c screened methyl orange. (Try writing with both positive and negative 'leads'.)

Extension 2

Can you find out which end of a battery is which?

Figure 5.10b shows a piece of red litmus paper wetted with sodium chloride solution. Leads from a battery touch it. Where lead 1 touches the paper, a white spot results. Where lead 2 touches the paper, a blue spot shows. Say which is the positive terminal of the battery, A or B.

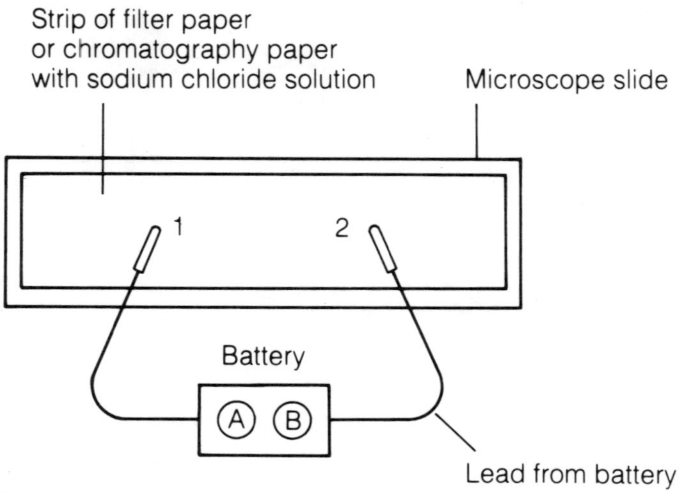

Fig. 5.10(b) Testing a battery

TOPIC 6 Formulas

Activity 6.1
Formula cards

1 Cut pieces of card to the following sizes:
size 1: 3 cm × 3 cm: 36 cards,
size 2: 3 cm × 6 cm: 32 cards,
size 3: 3 cm × 9 cm: 12 cards.

2 Colour half the cards red and half blue.

3 Write symbols of atoms and formulas of groups on the cards as listed here.

Size 1 red		*Size 2 red*		*Size 3 red*	
Na^+	(3 cards)	Zn^{2+}	(2 cards)	Al^{3+}	(2 cards)
K^+	(3 cards)	Cu^{2+}	(2 cards)	Fe^{3+}	(2 cards)
NH_4^+	(4 cards)	Ba^{2+}	(2 cards)	Cr^{3+}	(2 cards)
H^+	(4 cards)	Fe^{2+}	(2 cards)		
Ag^+	(3 cards)	Sn^{2+}	(2 cards)		
		Ca^{2+}	(2 cards)		
		Pb^{2+}	(2 cards)		
		Mg^{2+}	(2 cards)		

Size 1 blue		*Size 2 blue*		*Size 3 blue*	
Cl^-	(3 cards)	SO_3^{2-}	(2 cards)	PO_4^{3-}	(2 cards)
Br^-	(3 cards)	SO_4^{2-}	(4 cards)		
I^-	(2 cards)	CO_3^{2-}	(4 cards)		
NO_3^-	(2 cards)	O^{2-}	(2 cards)		
HCO_3^-	(2 cards)	S^{2-}	(2 cards)		
OH^-	(6 cards)				

4 By joining red cards and blue cards together, you can obtain the formula of a compound. An ionic compound contains both positive ions (red cards) and negative ions (blue cards). Join the tops and bottoms of the cards, but not the sides. A formula is square or rectangular, e.g.

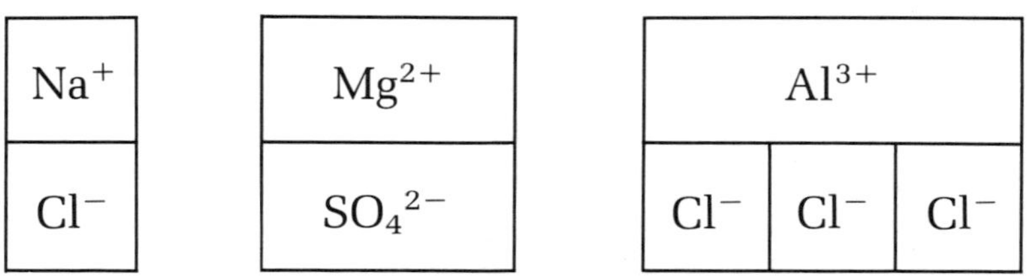

5 By putting cards together, arrive at the formulas for the following compounds. Write the formula of each compound before you go on to the next.

potassium iodide
ammonium nitrate
silver nitrate
magnesium sulphate
iron(II) iodide
aluminium iodide
aluminium oxide
iron(III) oxide
silver sulphide
chromium chloride
tin(II) chloride
sodium sulphite
potassium sulphate
silver sulphate
copper(II) sulphate
magnesium hydroxide
iron(III) bromide
aluminium hydroxide
aluminium sulphate
iron(III) sulphate
ammonium sulphate
chromium oxide
tin(II) sulphate
calcium sulphite
ammonium bromide
copper(II) carbonate
calcium hydroxide
magnesium hydrogencarbonate
barium hydroxide
sodium hydrogencarbonate
iron(III) hydroxide
iron(II) sulphate
sodium phosphate
chromium sulphate
calcium hydrogencarbonate
chromium hydroxide
any others you want to try.

Extension

Lego® bricks come in different sizes like these formula cards. They come in two colours, red and white, and they are made to stick together, as ions do. If you have some Lego® bricks at home, you can make yourself a set of formula bricks (Fig. 6.1), and assemble the formulas of ionic compounds as before.

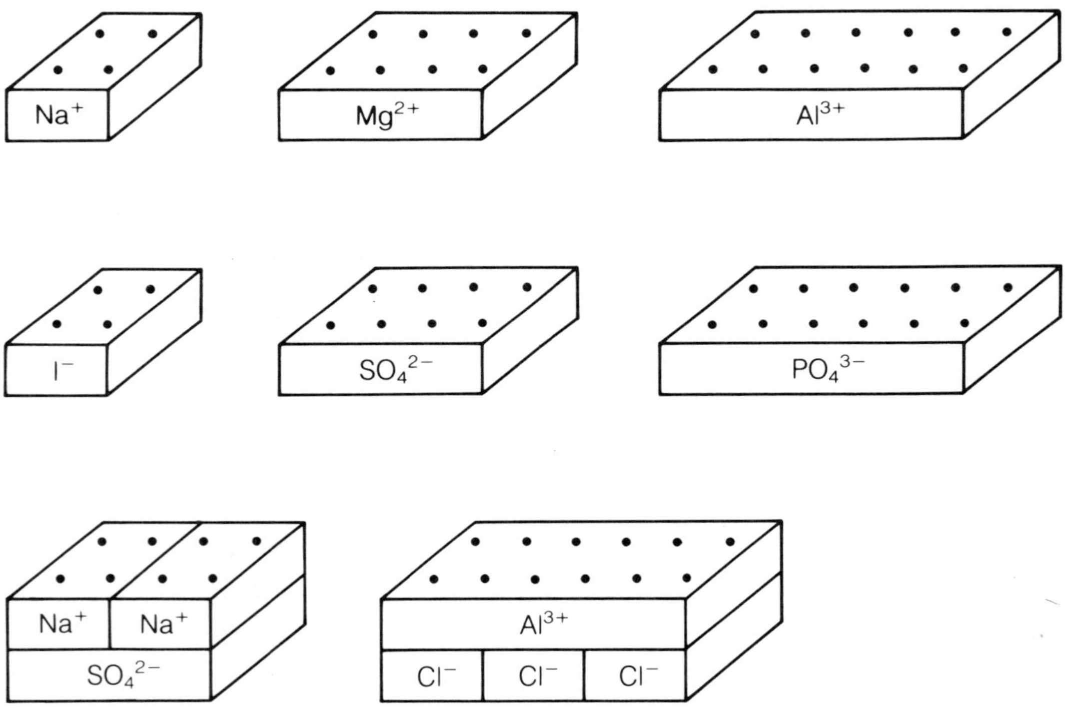

Fig. 6.1 Formula bricks

TOPIC 7 Acids, bases and salts

Experiment 7.1
Reactions of acids

1 Half fill a boiling tube with dilute hydrochloric acid. Add a few drops of litmus, and note the colour. Add sodium hydroxide solution dropwise from a teat pipette until the litmus changes colour. Note the new colour.

2 Evaporate the solution in the boiling tube to dryness by boiling gently over a Bunsen flame. What is left when all the water has evaporated?

3 Take a rack of test-tubes, each half full of dilute hydrochloric acid. To different test-tubes, add the metals magnesium, iron, zinc, copper and lead. If a gas is formed, collect it (see Fig. 7.1a). Test the gas with damp litmus paper and with a lighted splint (see Fig. 7.1b).

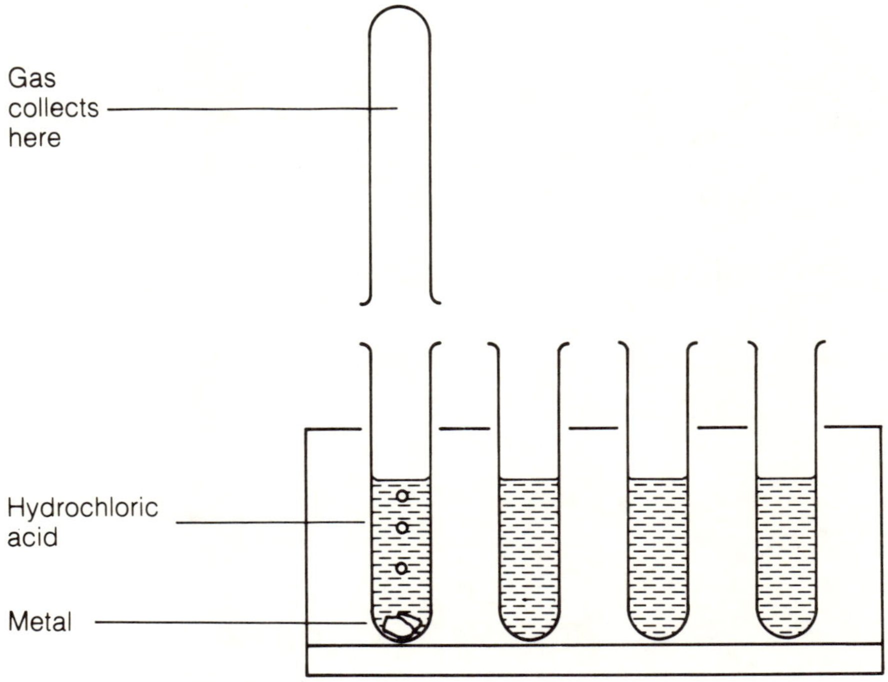

Fig. 7.1(a) Collecting the gas

4 Half fill a test-tube with dilute hydrochloric acid. Add a little copper(II) oxide. Warm gently. What change do you see? What chemical reaction is occurring?

5 Half fill a test-tube with dilute hydrochloric acid. Add a few crystals of sodium carbonate. If a gas is formed, test it with

- **a** damp litmus paper,
- **b** a lighted splint (see Fig. 7.1b),
- **c** limewater (see Fig. 7.1c).

6 Repeat all these tests with dilute sulphuric acid.

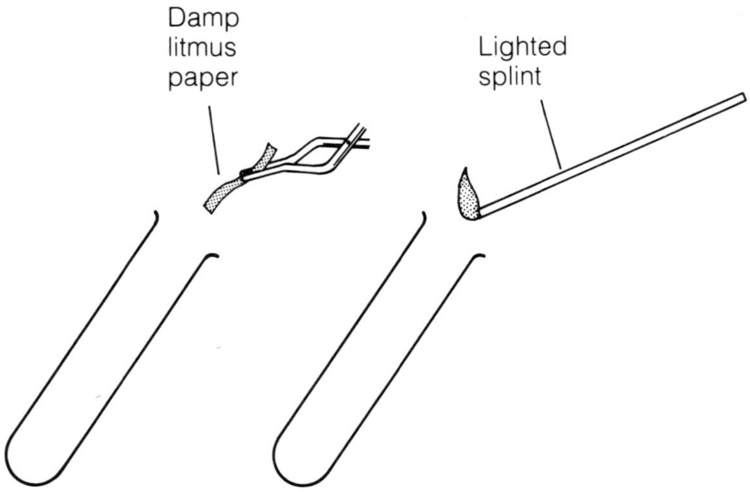

Fig. 7.1(b) Testing the gas

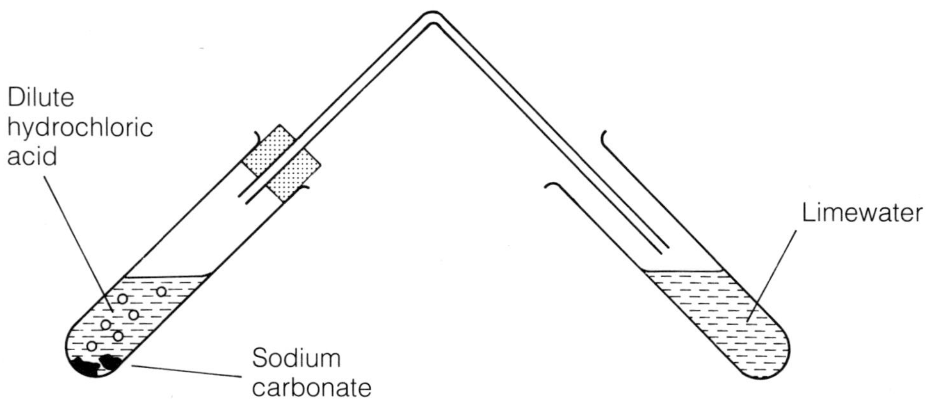

Fig. 7.1(c) Testing with limewater

7 Repeat again with a solution of ethanoic acid.

Experiment 7.2
Reactions of bases

Common laboratory bases are sodium hydroxide, ammonia, calcium hydroxide and calcium oxide. The first three mentioned are soluble; they are alkalis.

1 Half fill test-tubes with solutions of sodium hydroxide, ammonia and calcium hydroxide. Add litmus to each one, and note the colour.

2 Half fill test-tubes with solutions of the salts listed in the table. To each salt, add 5 cm^3 of sodium hydroxide solution.
Copy Table A, and make a note of what you see.

Salt	*Observation*	*Explanation: names of substances formed*
Copper(II) sulphate		
Iron(II) sulphate		
Iron(III) chloride		
Magnesium sulphate		

Copy and complete the word equations

copper(II) sulphate + sodium hydroxide → ———— + ————

iron(II) sulphate + sodium hydroxide → ———— + ————

iron(III) chloride + sodium hydroxide → ———— + ————

magnesium sulphate + sodium hydroxide → ———— + ————

Copy and complete the ionic equations

$Cu^{2+}(aq) + 2OH^-(aq) \rightarrow$ ———— (s)

$Fe^{2+}(aq) + 2OH^-(aq) \rightarrow$

$Fe^{3+}(aq) +$ ———— $\rightarrow$

$Mg^{2+}(aq) +$ ———— $\rightarrow$

Extension

1 Take three boiling tubes and fill one a quarter full with aluminium sulphate solution, another a quarter full with lead(II) nitrate solution and the third a quarter full with zinc sulphate solution.
To each tube, add sodium hydroxide solution in ½ cm^3 portions until you obtain a precipitate. Enter the appearance of the precipitates on a copy of Table B.

Table B Reactions of more salts with sodium hydroxide solution

Solution	*Result of adding sodium hydroxide*	*Result of adding an excess*
Aluminium sulphate		
Lead(II) nitrate		
Zinc sulphate		

Continue to add sodium hydroxide solution. Make a note of what you see. Try to explain your result. Then consult a textbook to see whether you are right.

2 Pour a little sodium hydroxide solution on to your thumb. Rub it between your finger and thumb. How would you describe the feel of sodium hydroxide?

3 Sodium hydroxide *saponifies* oil in your skin — reacts with the oil to make soap. You can do a small-scale saponification of oil or lard to make some soap. You need a concentrated solution of sodium hydroxide (200 g/l). You must use it with care; **wear safety glasses**. Sodium hydroxide is *caustic* (burning), and has the name *caustic soda*.

- **a** Put 25 g of olive oil (or lard) and 20 cm^3 of 20% sodium hydroxide solution into a conical flask.
- **b** Stand the conical flask on a tripod and gauze (Fig. 7.2a). Boil gently for 20 minutes. Use a very small flame: you do not want the mixture splashing up the sides of the flask. Holding the flask with tongs, swirl it from time to time. **Wear safety glasses**.
- **c** Pour the mixture into a beaker. Add 50 cm^3 of a saturated solution of sodium chloride. Boil and stir for 5 minutes.

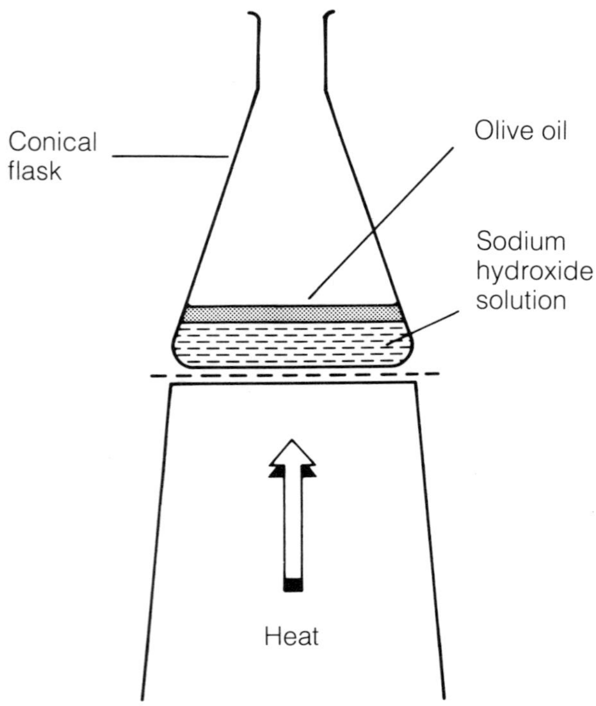

Fig. 7.2(a) Saponification

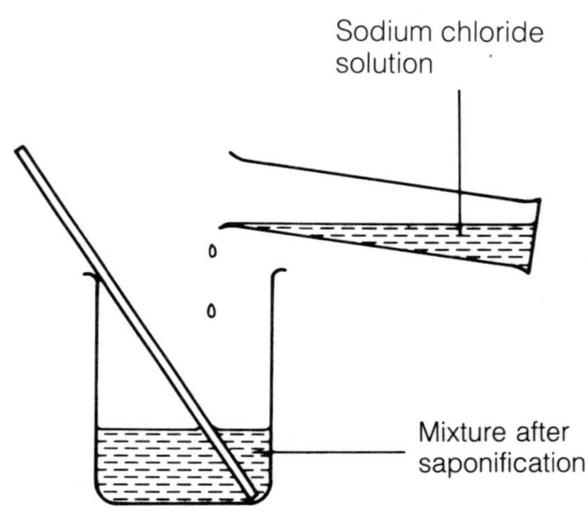

Fig. 7.2(b) Salting out

- **d** Let the mixture cool. The contents will separate into two layers. The upper solid layer is soap, and the bottom.layer is a mixture of glycerol and sodium chloride solution. Remove the layer of soap, and shape it into a bar. You can use it to wash your hands but not your face because it contains some caustic soda. (Commercial manufacturers take care to remove alkali from their soap.)

Experiment 7.3
What do you obtain when you neutralise vinegar with lime?

1 Warm 50 cm^3 of vinegar in a 250 cm^3 beaker.

2 Add one spatula measure of slaked lime. This is the lime you use in the garden. If all the slaked lime reacts, add some more.

3 Filter to remove the excess of slaked lime.

4 Evaporate the solution on a water bath until it starts to crystallise. (Do not use a Bunsen burner; this is hot enough to decompose the product.)

5 Vinegar is a dilute solution of ethanoic acid. Slaked lime is calcium hydroxide. It is a base used by farmers to neutralise acids in the soil. What are the crystals you have obtained? Write a word equation for the reaction:

ethanoic acid + calcium hydroxide → ———— + ————

6 Acids can compete. Ethanoic acid is a weak acid, and hydrochloric acid is a strong acid. What do you think will happen if you heat calcium ethanoate, which is a salt of a weak acid, with hydrochloric acid, which is a strong acid?

7 To find out, put one spatula measure of calcium ethanoate into a test-tube, and add 5 cm^3 of hydrochloric acid. Warm gently. Smell cautiously. Can you detect one of the products of the reaction? Copy and complete the word equation

calcium ethanoate + hydrochloric acid → ———— + ————

Experiment 7.4
Indicators

In this experiment you can revise the colours of some popular indicators.

The colours of some indicators

1 Copy Table A. Then carry out tests, and fill in the colours.

Table A Indicator colours

Indicator	*Colours*		
	Acidic solution	*Neutral solution*	*Alkaline solution*
Litmus			
Methyl orange			
Phenolphthalein			
Screened methyl orange			
Universal indicator			

Universal indicator

Universal indicator has more than one colour in acidic solution.

1 Make up a solution of universal indicator in tap water.

2 Use a measuring cylinder to pour 50 cm^3, 100 cm^3, 200 cm^3, 400 cm^3, 800 cm^3 and 1 litre of the solution into beakers.

3 You need a solution of ethanoic acid containing 3.0 g/l.

4 Add 10 cm^3 of the ethanoic acid solution to each beaker. Stir.

5 Record the colours in your copy of Table B. Compare the colours with the pH numbers on the universal indicator colour chart.

Table B Universal indicator in acidic solutions

	Volume of ethanoic acid in cm^3	*Volume of water in cm^3*	*Colour*	*pH*
Beaker 1	10	50		
Beaker 2	10	100		
Beaker 3	10	200		
Beaker 4	10	400		
Beaker 5	10	800		
Beaker 6	10	1000		

Which beaker has

a the lowest concentration of acid,

b the lowest pH?

6 Find the colour of universal indicator and the corresponding pH number for the following solutions and any others you want to try.
Make a copy of Table C, and enter your results.

Table C Universal indicator colours in various solutions

Solution	*Colour of universal indicator*	*pH*
Hydrochloric acid		
Ethanoic acid		
Tap water		
Distilled water		
Sodium hydroxide		
Ammonia		
Sodium carbonate		
Sodium hydrogencarbonate		
Sulphuric acid		
Sodium chloride		
Calcium hydroxide		

7 What is the colour of universal indicator in
a a strongly acidic solution,
b a weakly acidic solution,
c a neutral solution,
d a weakly alkaline solution,
e a strongly alkaline solution?

Experiment 7.5
Making zinc sulphate crystals

1 Take 50 cm^3 of dilute sulphuric acid in a Pyrex beaker. Add two pieces of granulated zinc, and warm (see Fig. 7.5a).

2 If all the zinc reacts, add more. Do not boil, but keep the reacting mixture warm. When no more bubbles of gas form, all the acid has been used up.

3 Filter to remove the excess of zinc (see Fig. 7.5b).

4 Evaporate the filtrate on a water bath. From time to time, dip in a glass rod, and hold it up to cool. When small crystals form on the glass rod, stop heating. Leave the solution to crystallise slowly.

5 Filter and dry the crystals of zinc sulphate.

6 What is the gas that is evolved? Write a word equation for the reaction. Write a balanced chemical equation.

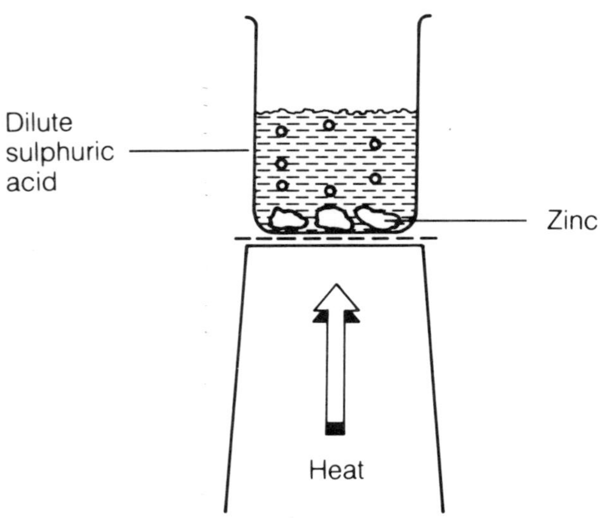

Fig. 7.5(a) Heating

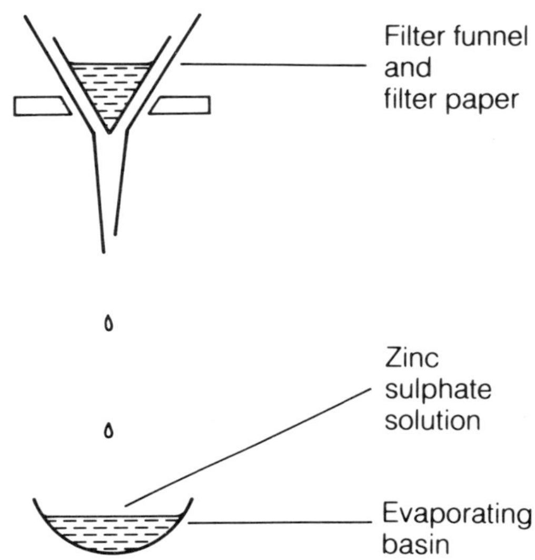

Fig. 7.5(b) Filtering

7 You must make sure that all the acid is used up in the reaction. If you do not do so, what will happen when you start to evaporate the solution of zinc sulphate? What did you do to ensure that all the acid would be used up?

8 In step **2**, you are told not to boil. What will happen if you heat too strongly or for too long in step **2**?

9 You can adapt this method to make other salts. Which starting materials would you use to make

- **a** magnesium sulphate,
- **b** nickel sulphate,
- **c** barium chloride?

Experiment 7.6
Making crystals of copper(II) sulphate-5-water

Copper(II) sulphate is an important fungicide. It is used for spraying grape vines and potato crops.

1 Take 50 cm^3 of dilute sulphuric acid in a Pyrex beaker. Add a spatula measure of copper(II) oxide. Warm, and stir with a glass rod (see Fig. 7.6a).

2 If all the copper(II) oxide reacts, add some more. To make sure that all the acid is used up, you must have an excess of copper(II) oxide. Test the solution from time to time with blue litmus paper. When the solution will no longer turn blue litmus paper red, all the acid has been used up. The reaction is then complete.

3 Filter to remove the excess of copper(II) oxide (see Fig. 7.6b).

4 Evaporate the filtrate on a water bath until crystallisation just begins. Then stop heating, and allow the solution to crystallise slowly.

5 Filter to obtain the crystals. You can keep the crystals by putting them into a stoppered specimen tube or by wrapping them in sellotape.

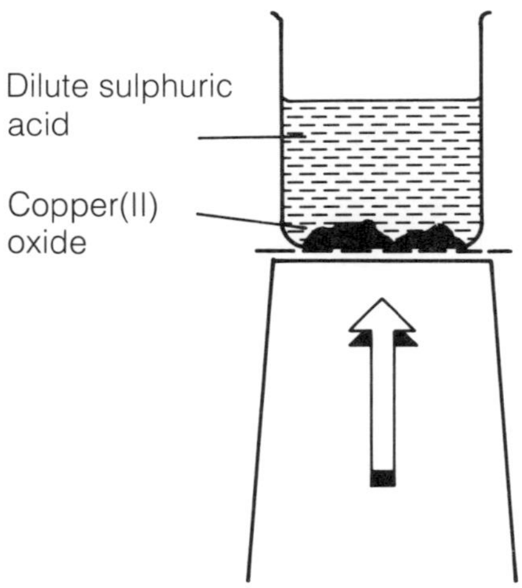

Fig. 7.6(a) Heating

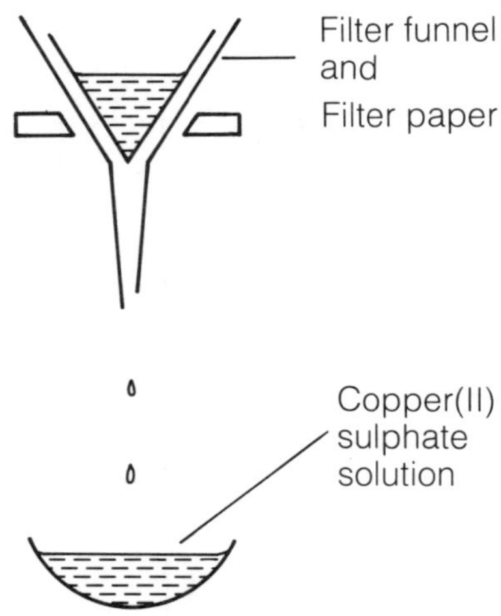

Fig. 7.6(b) Filtering

6 Why is it important to have no acid left at the end of the reaction?
How do you ensure that there is no acid left over?
How do you know when the reaction is finished?
Why do you not use the same method as you used in the preparation of zinc sulphate to find out when the reaction is finished?
Why do you not use the same method of preparation as you used for zinc sulphate, that is, the reaction between a metal and an acid?

7 Copy and complete the word equation

copper(II) oxide + sulphuric acid → ———— + water

Write a chemical equation for the reaction.

8 Which starting materials would you use to make
- **a** copper(II) chloride,
- **b** magnesium sulphate,
- **c** lead ethanoate?

Experiment 7.7
To heat crystals of copper(II) sulphate-5-water

1 Put some of the copper(II) sulphate-5-water crystals you have made into a side-arm boiling tube. Assemble the apparatus shown in Fig. 7.7.

2 Warm the crystals with a small flame.

3 Note the change in the thermometer reading.
Note the other changes which occur.

4 Answer the following questions.

- Why does the thermometer show a constant reading for a time?
- What happens during this time
  a to the crystals,
  b in the rest of the apparatus?
- What is the thermometer reading? What does this tell you?

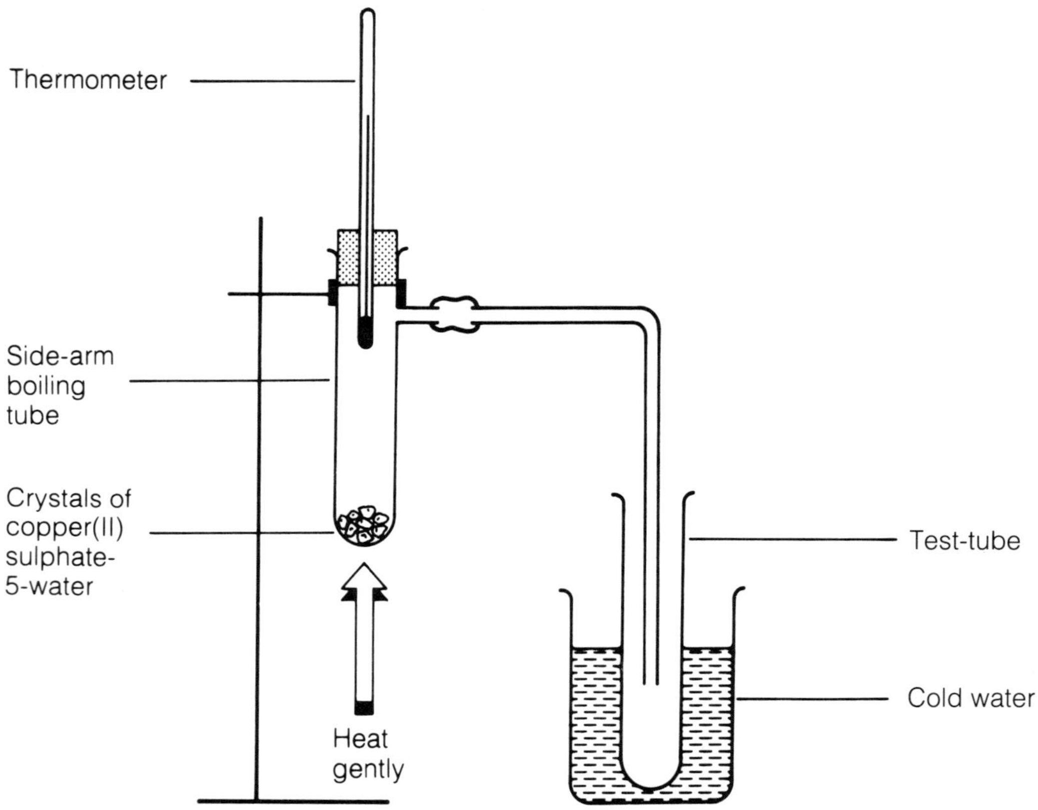

Fig. 7.7 Heating copper(II) sulphate-5-water

- What colour are the crystals
  a before heating,
  b after heating?
- What shape are the crystals
  a before heating,
  b after heating?
- Complete the word equation

  copper(II) sulphate crystals → ——— + ———
- Complete the chemical equation

  $CuSO_4{\cdot}5H_2O(s) \rightarrow$ ——— + ———

- Copy and complete the passage.

  Copper(II) sulphate-5-water crystals are ——— in colour and ——— in shape. When they are heated, they lose ——— to become ——— in colour and ——— in shape. The name of the product is ——— ——— ———. The name of the water in the crystals is ——— ——— ———. The water gives the crystals their ——— and ———.

5 What do you think will happen if you add water to the residue in the side-arm tube? Test to see whether you are right.

Experiment 7.8
Preparation of sodium sulphate crystals

This method depends on being able to add dilute sulphuric acid slowly and carefully to sodium hydroxide solution until the neutral salt, sodium sulphate, is formed. For this purpose, a burette is used (see Fig. 7.8). It is a glass tube, about 1 cm wide, with graduation marks along its length and a tap at the bottom. Watching the level of the acid in the burette pass over the graduation marks enables you to add acid in a measured way. The word for adding a liquid in a measured way is *titration*.

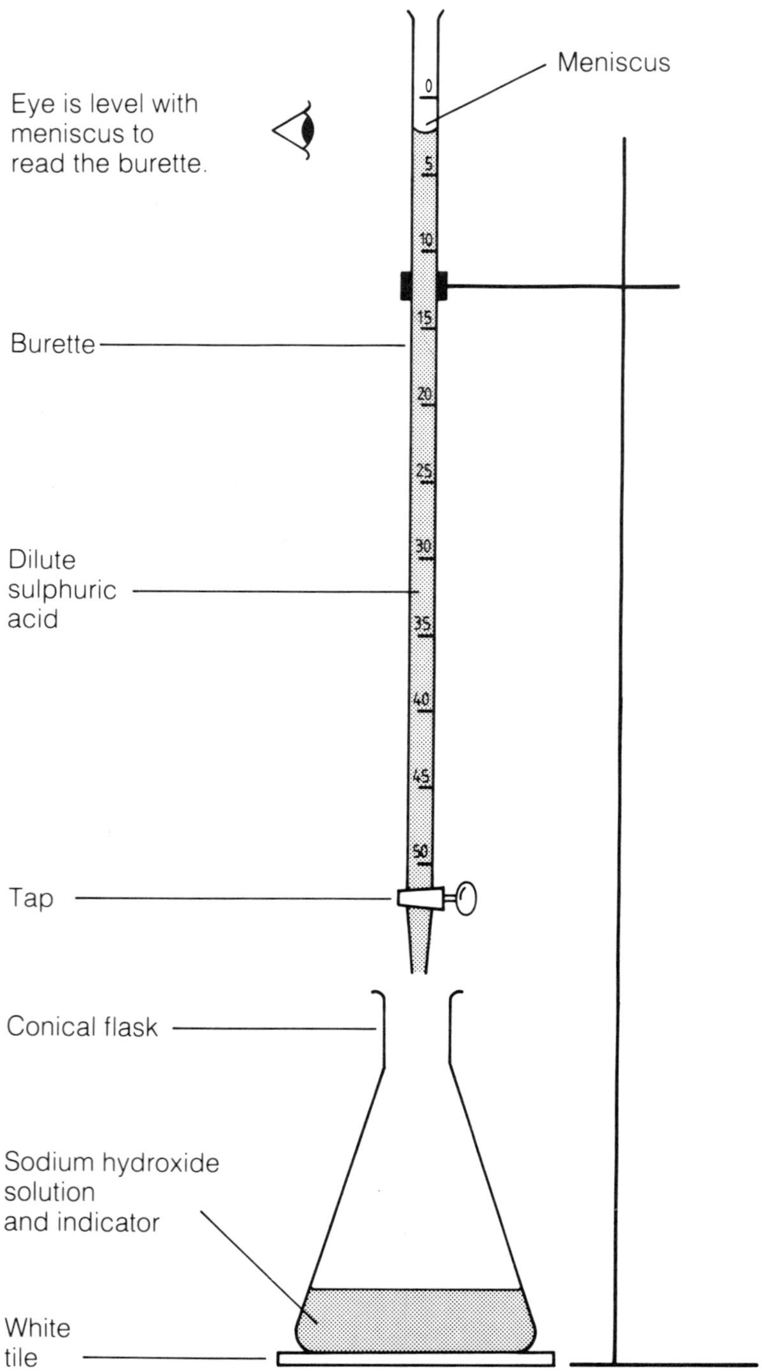

Fig. 7.8 Preparation of a sulphate by neutralisation

1 Set up the burette as shown in Fig. 7.8, with the tap closed. Wearing safety glasses, fill the burette through a funnel with dilute sulphuric acid. Open the tap to allow acid to fill the jet; then close the tap.

2 Read the burette. Figure 7.8 shows how to do this. The level of liquid is curved and is called a *meniscus*. Read the graduation mark at the bottom of the meniscus. Write down this volume, V_1 cm^3.

3 Using a pipette-filler, pipette 25 cm^3 of dilute sodium hydroxide solution into a conical flask. You must not pipette by mouth.

4 Add 3 drops of screened methyl orange indicator, and stand the conical flask on a white tile.

5 Open the tap to add dilute sulphuric acid slowly and carefully. Swirl the flask as you add acid. The indicator is yellow at the beginning and then turns green, and at the neutral point it suddenly turns *grey*. (If you add too much acid, the indicator turns purple, and you must repeat the titration.)

6 Read the burette, and write down this volume, V_2 cm^3.

7 The volume of sulphuric acid needed to neutralise 25 cm^3 of sodium hydroxide solution is $V_2 - V_1$ cm^3.

8 Wash out the conical flask. Pipette in 25 cm^3 of sodium hydroxide solution. **Do not add indicator**.

9 Refill the burette. From the burette, add $V_2 - V_1$ cm^3 of dilute sulphuric acid. This is the same volume as before; so you know that it is the correct volume to neutralise 25 cm^3 of sodium hydroxide solution.

10 Tip the solution in the conical flask into an evaporating basin, and evaporate until crystals form.

11 Why can the reaction between a metal and an acid not be used to make sodium salts?
In Experiments 7.5 and 7.6, you used an excess of one of the reactants. What method did you use to remove the excess? Why can you not use an excess of one reagent in this experiment?

Experiment 7.9
Preparation of insoluble salts

Preparation of lead(II) chromate

1 In a measuring cylinder, take 20 cm^3 of lead(II) nitrate solution. In a second measuring cylinder, take 20 cm^3 of potassium chromate solution.

2 Pour both solutions into a beaker. Stir.

3 Filter the precipitate of lead(II) chromate. Wash it with a little water while it is on the filter paper.

4 Turn it on to a watch glass and leave it to dry.

5 Write a word equation for the reaction.

Preparation of lead(II) iodide

Repeat steps **1–4** but this time use potassium iodide solution instead of potassium chromate solution. Write a word equation for the reaction.

A chemical garden

Water glass is a liquid which is used for coating paper and wood to seal the pores. It is sodium silicate.

1 Mix 50 cm^3 of water glass with 250 cm^3 of water. Stir thoroughly.

2 Use a ½ kg (1 lb) jam jar. Beakers are difficult to clean after this experiment. Pour the solution into the jar.

3 Drop in crystals of iron(II) sulphate, iron(III) chloride, copper(II) sulphate, calcium nitrate, manganese(II) sulphate, nickel sulphate, cobalt(II) chloride, chromium(III) sulphate and magnesium sulphate. After a few hours, you will have a chemical garden. The growths are insoluble silicates. Most metal silicates are insoluble. Copy and complete the word equations

iron(II) sulphate + sodium silicate → ——— ——— + ——— ———

copper(II) sulphate + sodium silicate → ——— ——— + ——— ———

cobalt(II) chloride + sodium silicate → ——— ——— + ——— ———

Experiment 7.10
Comparing indigestion tablets

You need a selection of 'acid indigestion' tablets, e.g. Rennie®, Bisodol®, Milk of Magnesia® (solid and liquid) and others. You need either methyl orange or bromophenol blue indicator and a 2.0 mol/l solution of hydrochloric acid in a burette.

1 Place one tablet on a piece of paper, and crush it with a spatula. Fold the paper, and tip the powdered tablet carefully into a 100 cm^3 flask. Add about 25 cm^3 of water. Add 5 drops of indicator. You can use methyl orange (which changes from yellow to red) or bromophenol blue (which changes from blue to yellow).

2 Fill a burette with bench hydrochloric acid (2 mol/l). Read the burette.

3 Add hydrochloric acid from the burette. Shake well after each addition. Stop when methyl orange changes to a permanent red colour (or when bromophenol blue changes to a permanent yellow colour). The insolubility of some of the tablets makes the neutralisation take place slowly. Shaking well as you titrate assists neutralisation.

4 Repeat the procedure for other tablets and liquids. Tabulate your results. Compare the price of the tablets with the volume of acid needed to neutralise one tablet. Calculate the value (in cm^3/p) of

$$\frac{\text{Number of tablets in packet}}{\text{Price of packet}} \times \begin{array}{l}\text{Volume of bench hydrochloric acid}\\\text{needed to neutralise one tablet}\end{array}$$

Neutralising indigestion tablets

Name of tablet	*Number in packet*	*Price of packet in p*	*Chemicals in tablet*	*Volume of acid needed in cm^3*	*Speed of titration (slow/ fast/ medium)*	*Calculated value in cm^3/p*
Rennie®			Magnesium carbonate			
etc.						

This will give you a means of ranking the different brands. Enter the calculated value in the table. Is there anything else you would consider before making your choice of 'best buy'?

Experiment 7.11
A long experiment to do at home

Bacteria change lactose, the sugar in milk, into lactic acid. Lactic acid gives sour milk its acidic taste. Like other acids, it is neutralised by alkalis.

1 Dissolve some sodium hydrogencarbonate in 250 cm^3 of milk. Allow the milk to stand in a warm place. Have 250 cm^3 of untreated milk standing beside it. Which goes sour first?

2 Add some sodium hydrogencarbonate ('bicarbonate of soda') to the sour milk. Is the taste altered?

TOPIC 8 Air

Experiment 8.1
Burning elements in oxygen

1 Fill four boiling tubes with oxygen from a cylinder. As oxygen is insoluble in water, it can be collected as shown in Fig. 8.1a. The cylinder valve should be only slightly open to release a slow stream of gas. When a boiling tube is full, cork the open end while it is still under water. Then take the boiling tube out of the trough, and stand it in a rack.

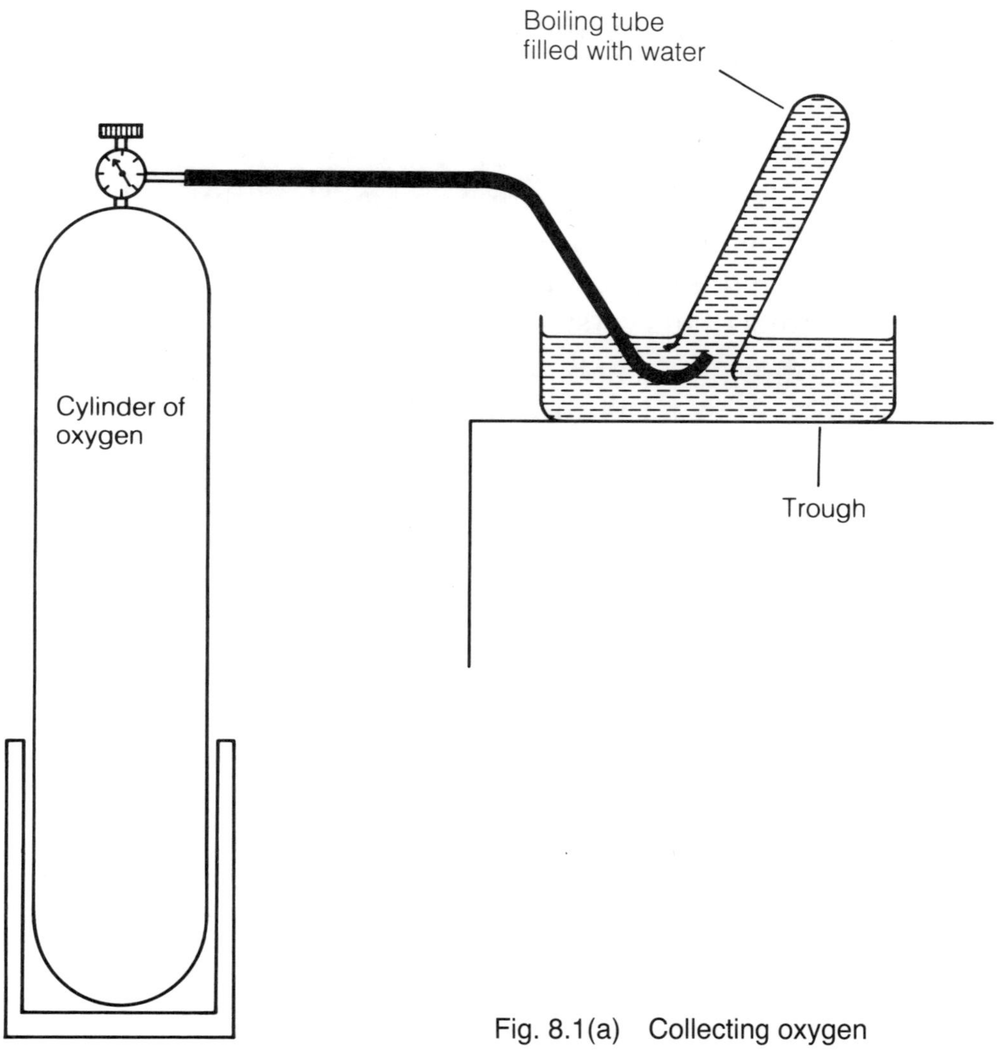

Fig. 8.1(a) Collecting oxygen

2 Wrap some iron wool round a combustion spoon. Heat it in a Bunsen flame to red heat. Then uncork one of the boiling tubes of oxygen, and lower the combustion spoon into it. Keep the boiling tube in the rack; do not hold it in your hand (Fig. 8.1b). Note the way in which the iron burns. After 10 seconds, remove the combustion spoon, and replace the cork. Add three drops of universal indicator and shake.

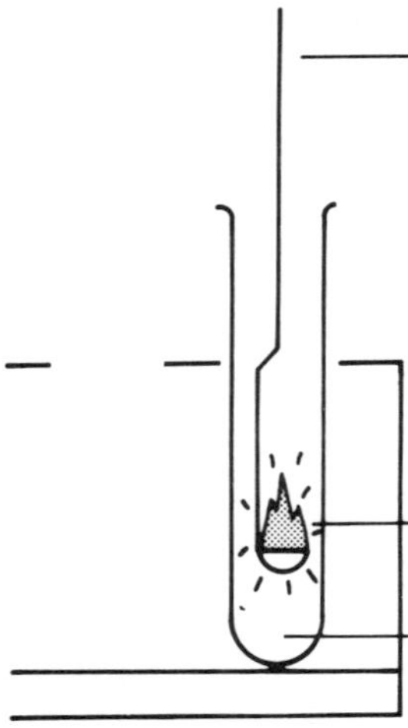

Fig. 8.1(b) Burning an element in oxygen

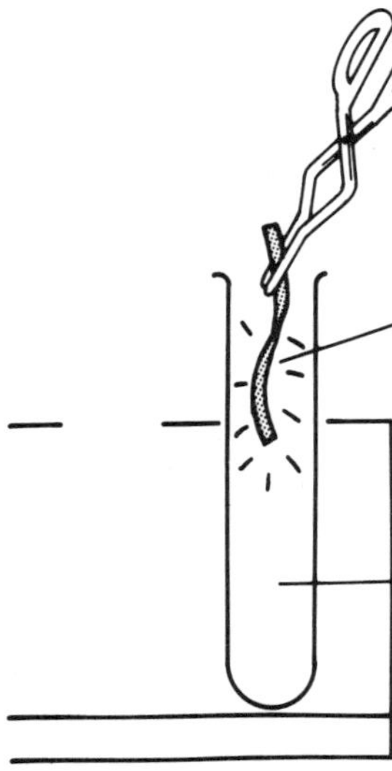

Fig. 8.1(c) Burning magnesium in oxygen

3 Grip a 5 cm length of magnesium ribbon in tongs. Hold it in a Bunsen flame until it starts to burn. Then quickly lower the burning magnesium into a boiling tube of oxygen (Fig. 8.1c). After it stops burning, replace the cork. Add 3 drops of universal indicator, and shake.

4 Place some charcoal on a combustion spoon. Hold it in a Bunsen flame until it is red hot. Then quickly insert the combustion spoon into a boiling tube of oxygen. After 10 seconds, remove the combustion spoon. Replace the cork. Add 3 drops of universal indicator solution, and shake.

5 Place some sulphur on a combustion spoon. Heat it until the sulphur burns. Then quickly transfer it into a boiling tube of oxygen. Note how the sulphur burns. After 10 seconds, remove the combustion spoon, and replace the cork. Add 3 drops of universal indicator and shake.

6 Ask your teacher to demonstrate the combustion of sodium, calcium and phosphorus.

7 Make out a copy of the table, and fill in your results.

Table Combustion of elements in oxygen

Element	*Metallic or non-metallic*	*Appearance on burning*	*Appearance of product*	*Colour of universal indicator*	*Is solution acidic or alkaline?*	*Is product an acid or a base?*
Iron						
Magnesium						
Carbon						
Sulphur						
Sodium						
Calcium						
Phosphorus						

8 On the basis of your results, copy and complete the following passage. Elements burn in oxygen faster/more slowly than they do in air. Metallic elements burn in oxygen to form ——— oxides; non-metallic elements burn in oxygen to form ——— oxides.

Note: If you do not have a cylinder of oxygen, you can make some. Figure 8.1d shows an apparatus you can use. If you have gas jars, use them instead of boiling tubes.

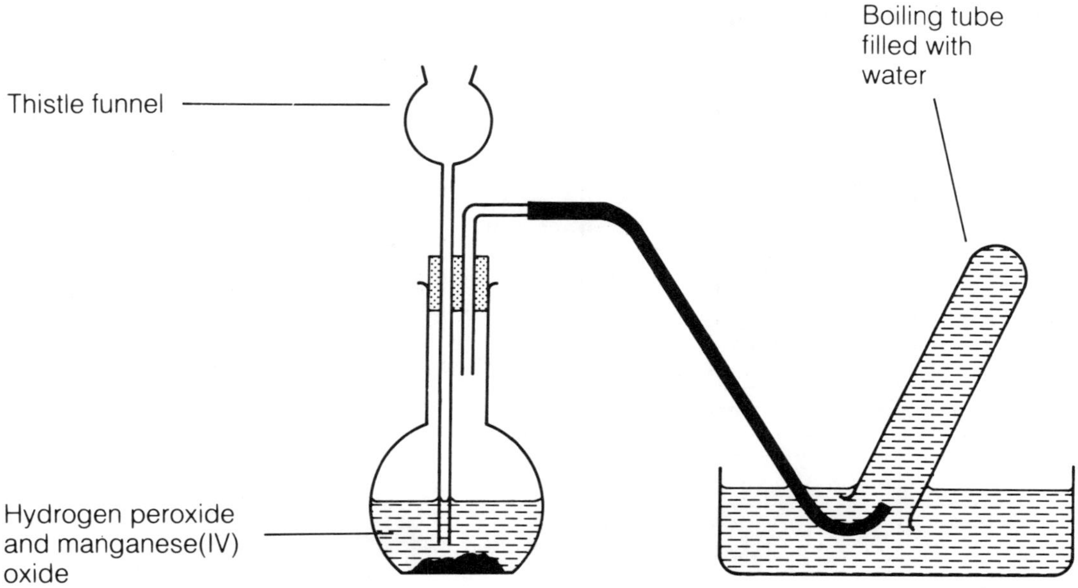

Figure 8.1(d) Apparatus for making oxygen

Experiment 8.2
Under what conditions does iron rust?

Take nine clean iron nails.

1 Set up a rack of test-tubes (see Fig. 8.2). Into the first, put an iron nail. Cover the nail with tap water. Cork the test-tube, and label it 'tap water'.

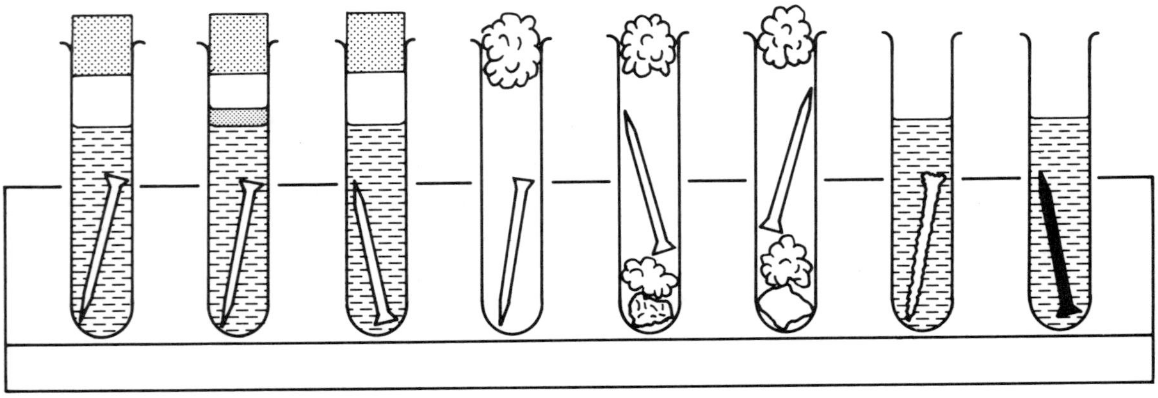

Fig. 8.2 Rusting experiments

2 Boil some tap water in a boiling tube. Keep the water on the boil for 5 minutes. This will drive out dissolved air. Tip the water into one of the test-tubes in the rack. Add a nail, and boil again. Pour in 3 cm^3 of paraffin oil. This will form a thin layer on top of the water. Cork the test-tube, and label it 'boiled-out water and oil'.

3 Repeat step **2**, but omit the paraffin oil. Label the test-tube 'boiled-out water'.

4 Place a nail in a dry test-tube. Stopper the tube with cotton wool.

5 Put a lump of anhydrous calcium chloride into a test-tube. Cover it with cotton wool. Place a nail on top of the cotton wool. Stopper the tube with cotton wool and label it.

6 Put a lump of soda–lime in a test-tube. Cover it with cotton wool. Drop in a nail. Stopper the tube with cotton wool, and label it.

7 Cover a nail in grease. Put it into a test-tube full of water.

8 Paint a nail. Let it dry. Then put it into a test-tube full of water.

9 Leave the rack of test-tubes for 1 week. Examine the nails for signs of rusting.

10 After 1 month, look at the nails again.
What gas does anhydrous calcium chloride remove from the air?
What gas does soda–lime take from the air?
What purpose does the layer of paraffin oil serve?
List the substances which must be present before iron will rust.

Extension ***Does iron increase in mass when it rusts or decrease or stay the same?***

Design an experiment to find out. You are supplied with iron wool, iron nails, petri dishes, test-tubes, beakers, an oven, a balance and other normal laboratory apparatus. Think carefully what weighings you will need to make. Write an outline of the experiment you have planned. If your teacher approves it, proceed to carry it out.

Experiment 8.3
What percentage of the air is used in rusting?

1 Take a 100 cm^3 graduated tube. You need to fill it with enough water to leave 100 cm^3 of air inside. Add some water, put your thumb over the open end, and turn the tube upside down. Let water trickle out until you have 100 cm^3 of air in the graduated tube.

2 Transfer the tube to a deep container (Fig. 8.3a). Move it until the levels of water inside and outside the tube are the same. The pressure of air inside the graduated tube is now the same as that outside, that is, atmospheric pressure. Read the volume of air.

3 Put your thumb over the end of the tube. Turn it open end upwards. Push a wad of iron wool to the end of the tube.

4 Invert the tube in a beaker of water. Leave it to stand for a week (Fig. 8.3b).

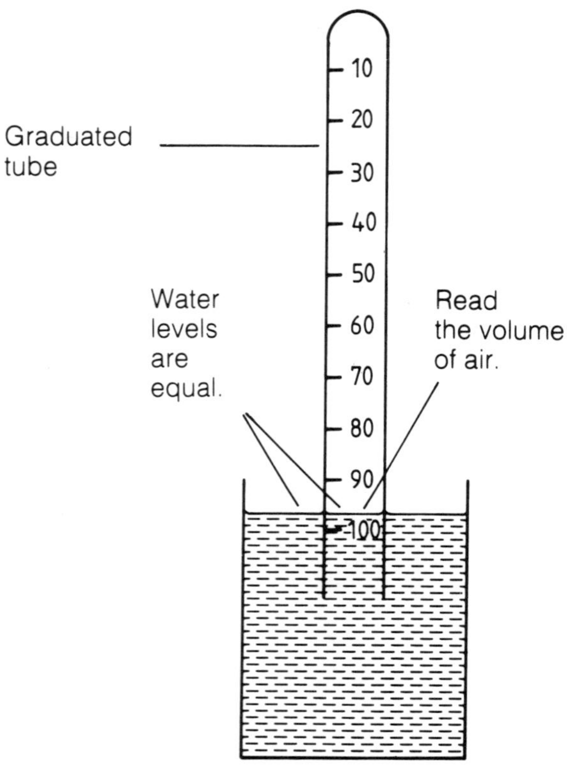

Fig. 8.3(a) Measuring the volume of air

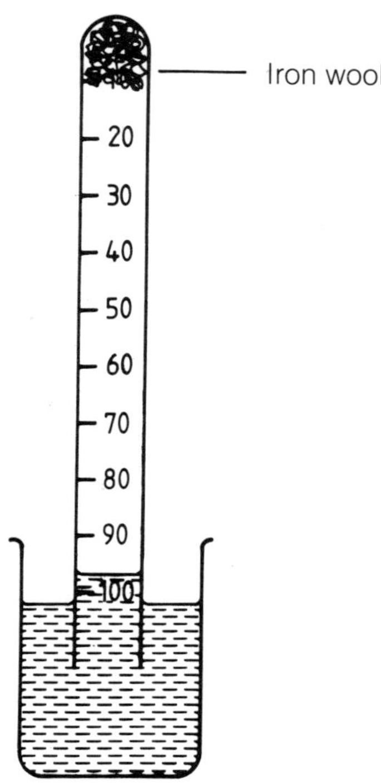

Fig. 8.3(b) Leaving iron to rust

5 With your thumb over the open end, transfer the graduated tube to a deep container. Adjust its position until the levels of water inside and outside the tube are the same. The air inside the tube is now at atmospheric pressure. Read its volume.

6 Test the gas inside the tube with a lighted splint. Is it still air?

7 List your results.
Volume of air to start with = 100 cm^3
Volume of air after iron has rusted = v cm^3
Volume of air used up = $(100 - v)$ cm^3
Percentage of air used up = $(100 - v)\%$

Experiment 8.4
Does sulphur dioxide affect plant growth?

You may have read that many people are worried about the vast quantities of sulphur dioxide which are given off by the chimneys of factories and power stations. In this experiment, you are asked to find out whether sulphur dioxide affects the growth of plants.

You are supplied with a packet of cress seeds, empty yoghurt cartons, aluminium foil, empty margarine tubs, crystallising dishes, beakers, plastic bags with bag ties, labels, plastic food boxes, sand, soil and cotton wool. You can make sulphur dioxide from sodium metabisulphite and water. These substances react slowly to release a gradual stream of sulphur dioxide.

Can you plan an experiment to find out whether cress will grow better in air or in air that contains sulphur dioxide?

You have to decide how you are going to plant the cress seed. You know that you can get a supply of sulphur dioxide from sodium metabisulphite and water. As in all investigations, you must have a control experiment.

Experiment 8.5
Some reactions of hydrogen peroxide

You need grass, green leaves, flower petals, hydrogen peroxide solution (4 volume).

1 Take a rack of test-tubes. Into one put some grass, into another put a green leaf, and into the rest put flower petals of different colours.

2 Add 5 cm^3 of hydrogen peroxide solution to each test-tube. Leave the test-tubes to stand for 15–30 minutes. Look at them from time to time. What action does hydrogen peroxide have on the specimens? (Look carefully; do not assume that the result will be the same in every case.)

3 Take enough manganese(IV) oxide to cover the tip of a spatula, and place it in a test-tube. Add 5 cm^3 of hydrogen peroxide solution (4 volume). Test the gas evolved with

- **a** a glowing splint,
- **b** a lighted splint,
- **c** damp blue litmus paper.

4 Pour 5 cm^3 of dilute hydrochloric acid into a test-tube. Add 5 cm^3 of hydrogen peroxide solution. If the reaction is very slow, warm gently. A gas is produced. Test it with

- **a** a lighted splint,
- **b** damp blue litmus paper.

What do you think the gas is?

5 Take 5 cm^3 of potassium iodide solution. Add 5 cm^3 of bench sulphuric acid and 5 cm^3 of hydrogen peroxide solution.
Observe what happens. What substance has been formed?
Add 1 cm^3 of hexane (or trichloroethane). Explain what you see.

6 Take 5 cm^3 of potassium bromide solution. Add 5 cm^3 of bench sulphuric acid and 5 cm^3 of hydrogen peroxide solution.
Observe what happens. What substance has been formed?
Add 1 cm^3 of hexane (or trichloroethane). Explain what you see.

Extension *Invisible writing and hydrogen peroxide*

Gallic acid is properly called 3,4,5-trihydroxybenzoic acid.

1 Make a solution of gallic acid.

2 Make a solution of iron(II) sulphate. Add a little dilute sulphuric acid. It is important to use a freshly made solution of iron(II) sulphate.

3 Add the two solutions, and mix them by shaking.

4 Write on a sheet of paper with the solution. Use a brush if you have one; otherwise a teat pipette or a glass rod will do. Let the writing dry, either by leaving it at room temperature or by holding it near a Bunsen flame. The writing will be invisible.

5 Make up a dilute solution of hydrogen peroxide. It must be '½ volume' hydrogen peroxide. Wet a paper towel with the dilute hydrogen peroxide, and use it to dab the paper you wrote on.

6 Iron(II) compounds are oxidised by hydrogen peroxide to iron(III) compounds. What can you say about the difference between iron(II) gallate and iron(III) gallate?

TOPIC 9 Water

Experiment 9.1
Which ions make water hard?

Distilled water is soft; it lathers well with soap without forming scum. When some salts are dissolved in water, the water is hard; it produces a small amount of lather and much scum. In this experiment, you can find out which ions make water hard.

1 Make a mark half-way up a test-tube. You can use a marking pen or a rubber band. Pour distilled water into the test-tube up to the mark.

2 Add 1 cm^3 of soap solution. You can do this from a graduated 1 cm^3 pipette. If you have no graduated pipette, take a teat pipette, and make a mark on it with a glass knife at about 1 cm^3. You can then add the same volume of soap solution each time.

3 Shake the test-tube for 10 seconds. Measure the height (in mm) of the lather on top of the water.

4 Make a copy of the table. Repeat the test for each of the solutions listed in the table. Enter the results you obtain.

Table Lather formed in solutions of salts

Solution	*Ions present*	*Height of lather in mm*
Sodium chloride		
Sodium sulphate		
Sodium carbonate		
Sodium nitrate		
Calcium chloride		
Calcium nitrate		
Potassium chloride		
Potassium nitrate		
Magnesium chloride		
Magnesium sulphate		
Magnesium nitrate		

5 You will easily see which *salts* make water hard. If you list the ions present, you should see which *ions* are responsible for making water hard.

Experiment 9.2
Methods of softening tap water

You can assess how hard a sample of water is by finding the volume of soap solution needed to give a lasting lather in the water. A sample of hard water needs more soap than soft water.

1 Fill a burette with soap solution. Read the burette.

2 Tap water. Measure 25 cm^3 of tap water in a measuring cylinder. Pour it into a conical flask. Add soap solution slowly from a burette, drop by drop, swirling the conical flask all the time. When you get a lather which lasts for 10 seconds, stop. Read the burette. Write down the volume of soap solution you use.

3 Boiled water. Boil some tap water in a beaker for 5 minutes. Let it cool; then repeat step **2** with the boiled water.

4 Distilled water. Repeat step **2**, using distilled water.

5 Washing soda. This time, dissolve a crystal of washing soda (sodium carbonate-10-water) in 25 cm^3 of tap water. Then repeat step **2**.

6 Washing powder. Repeat step **5** using the brand of washing powder which you use at home.

7 Make a list of your results. Which method of softening worked best for you? Did boiling the tap water soften it? If it did, what does this prove about the water in your area?

Extension

1 In parts of the country which have hard water, some houses have water softeners which use *ion exchange resins*, such as Permutit®. Tap water runs through a cylinder containing Permutit® and leaves the bottom of the cylinder softened. Can you make a model of a domestic water softener? You are supplied with Permutit® and lengths of glass tubing of width 4 cm, 2 cm, 1 cm and ½ cm. You have rubber bungs, corks, cork-borers, rubber tubing, glass wool, cotton wool, screw clips, glass taps and any other normal laboratory apparatus.

2 Sketch an apparatus which you think will soften water. Ask your teacher to approve your plan, and then assemble the model water softener.

3 Test the water which comes out of your model water softener as you did in step **2** of the main experiment.

4 Make a concentrated solution of cobalt sulphate. The pink colour of the solution is due to the cobalt ions. Pour 10 cm^3 into the Permutit® water softener you have made. When 5 cm^3 of the solution have passed through the column, pour two 10 cm^3 portions of distilled water through the column. What happens to the cobalt ions?
Now make a concentrated solution of sodium chloride. Pour it into the column. From what you observe, say what happens to the cobalt ions. Explain why this change happens.

Experiment 9.3
Methods for getting rid of a film of oil on water

1 Pour some brine (sodium chloride solution) into six plastic cups. With a teat pipette, add 1 cm^3 of crude oil to each. This resembles an oil spill at sea. Test the six cups as follows.

2 Add undiluted liquid detergent, drop by drop.

3 Add Polyclens®.

4 Sprinkle the oil with sand.

5 Sprinkle the oil with plaster of Paris.

6 Sprinkle the oil with sawdust.

7 Sprinkle the oil with chips of polystyrene foam.

8 Try out any more ideas you have for treating the oil. **Do not try to burn the oil**. This is dangerous because it will make the water boil and spit out burning drops of oil.

9 On the basis of your experiments, do you think that dispersing the oil with a detergent or absorbing the oil with a solid absorbent is to be recommended?

Experiment 9.4
Comparing soaps with detergents

Surface tension

1 Half fill a 250 cm^3 beaker with water. Sprinkle some powdered charcoal on the surface. Add some detergent solution.

2 Repeat, but this time add soap solution to the water.

3 With a teat pipette, let drops of water fall on to a flat piece of cloth.

4 Repeat step **3**, using water to which you have added soap solution.

5 Repeat step **3**, using water to which you have added detergent.

6 Explain what you have seen in terms of the surface tension of water, soap solution and detergent solution.

Emulsifying oil

1 Half fill a boiling tube with water. Add 2 cm^3 of a coloured oil, stopper the tube, and shake.

2 Add 2 cm^3 of soap solution, and shake.

3 Repeat step **1**, then add 2 cm^3 of detergent, and shake.

4 In which boiling tube have oil and water emulsified most?

Obtaining a lather with hard water

1 Make a solution of 1 g of magnesium sulphate crystals in 100 cm^3 of water.

2 Take 20 cm^3 of the solution in a boiling tube. Add 2 cm^3 of soap solution, stopper the tube, and shake.

3 To 20 cm^3 of the solution in a boiling tube, add 2 cm^3 of detergent solution, and shake.

4 Repeat steps **2** and **3** using distilled water.

5 Make a table of your results.

Table Appearance of lather

	Hard water	*Distilled water*
Soap solution		
Detergent		

Removing grease from cloth

1 Make some grease from powdered carbon and oil. Smear it on to three pieces of cloth. You should use the same type of cloth for all the tests.

2 Take three 250 cm^3 beakers. Into each, put 200 cm^3 of tap water. In one beaker, dissolve 5 g of liquid detergent. In the second, dissolve 5 g of solid detergent, and, in the third, dissolve 5 g of soap flakes.

3 Place one piece of stained cloth in each beaker. Stir each with a glass rod for 10 minutes.

4 Remove the pieces of cloth. Examine them to see which cleaning solution has done the best job.

Extension

Do further experiments to find out which temperature is best for removing grease stains. Your washing machine calls 40 °C a warm wash, 60 °C a hot wash and 85 °C a very hot wash. You may like to try these three temperatures. Write a brief description of the tests you plan. Show it to your teacher, and, if he or she approves, carry out your plan.

On the basis of your experiments, recommend a method of removing grease stains from cloth.

Experiment 9.5
Comparing different detergents

You need a selection of detergents.

Is it a soap or a soapless detergent?

1 Shake some detergent with cold water. Note whether the detergent dissolves easily. Note whether foam or lather is formed. Add some dilute hydrochloric acid, and note whether the lather stays.

A soap does not dissolve easily in cold water, and its lather is destroyed by hydrochloric acid.

2 Put the detergent into a boiling tube containing some ethanol. Stand the boiling tube in a beaker of hot water. Soap will dissolve and colour the ethanol. If you cool the solution, it will set to a jelly.

What is the pH of a solution of the detergent?

1 Make solutions of the detergents you have collected. Use 1 g in 100 cm^3 of distilled water.

2 To 10 cm^3 of a detergent solution, add a few drops of universal indicator. Compare the colour with the universal indicator chart. Write down the pH of the detergent solution.

Does the detergent contain bleach?

1 Lay a piece of starch–iodide paper on a white tile. Add a little dilute hydrochloric acid. Add some detergent solution. If the paper turns blue, bleach is present.
(This test is explained in Topic 15.)

How much water does the detergent contain?

1 Weigh a small metal dish. Write down the mass m_1.

2 Add about 5 g of detergent, and reweigh. Write down the mass m_2.

3 Put the metal dish into an oven at 105 °C. Leave it overnight.

4 Reweigh the metal dish and detergent. Record the mass m_3.

5 Reheat for an hour, cool and reweigh. Heating and weighing should be repeated until there is no further loss in mass. Record the mass m_4.

6 Find the mass of detergent $(m_2 - m_1)$.
Find the mass of water in it $(m_2 - m_4)$.
Calculate the percentage by mass of water in the detergent:

$$\text{percentage of water} = \frac{m_2 - m_4}{m_2 - m_1} \times 100\%$$

7 Repeat the procedure with other detergents. Tabulate your results.

Table Detergents

Detergent	*Price in p*	*Mass in packet in g*	*Price in p/100 g*	*Soap or soapless*	*pH of solution*	*Bleach present?*	*Percentage of water*
Kleeno							
Wyto							

Arrange the detergents in order of mass of anhydrous detergent/price. What else would you have to know before you could say which was the best value for money?

Extension *The foaming power of detergents*

Design an experiment to compare the foaming powers of different detergents.

a How will you measure the quantity of foam produced?
b Will you measure anything apart from the *quantity* of foam produced?
c What factors will you have to keep the same to make a fair comparison?
Plan your procedure. Obtain a variety of detergents, and carry out your tests. Find out whether you obtain the same comparison under different conditions.
d Is the quantity of foam produced the only thing that matters? How could you find out whether the detergent that makes the most foam does the best cleaning job?

Take your best foamer and your worst foamer. Plan a test to see which of the two cleans better. Carry out the test you have planned.

Experiment 9.6
Which work better: enzyme detergents or non-enzyme detergents?

Many detergents contain enzymes. Enzymes are substances which *catalyse* biological reactions, that is, chemical reactions which take place in plants or animals. A catalyst increases the speed at which a reaction takes place.
You need

a an enzyme detergent and a non-enzyme detergent (1% solutions),
b some white cotton cloth and some of the substances which often stain clothing, such as coffee, tea, fruit juice, egg yolk, beer, grass and blood. (You can ask the butcher for some blood.)

You have to decide

a how to stain pieces of cloth with each substance,
b whether you will let the stain dry,
c a fair method of comparing the washing action of enzyme detergent and non-enzyme detergent on the same stain,
d how long you will wash the stained cloth,
e a method of labelling each experiment so that you know which stain and which detergent you are using,
f a method of assessing which washing powder did the better job.

When you have finished, tabulate your results.

Table Appearance of stains after soaking in detergents

Stain	*Enzyme detergent*	*Non-enzyme detergent*
Tea		
Coffee		
etc.		

Write a summary of your results.

Extension 1 *Temperature*

Choose a stain for which enzyme detergent worked well.

Repeat the comparison of enzyme detergent and non-enzyme detergent with this stain at 50 °C and again at 90 °C. (You should find a difference. Can you explain it? Enzymes are proteins. You will have to find out something about proteins before you can explain the difference.)

Extension 2 *Salt*

Before the days of enzyme detergents, people used to soak stains in a solution of common salt (sodium chloride). Do a test to compare the effectiveness of sodium chloride and enzyme detergent in dealing with some of the staining substances you have.

Experiment 9.7
Should you use the same detergent for every type of fabric?

1 You need a selection of fabrics: wool, cotton, nylon, Terylene, polyester, Crimplene, rayon etc. They should be white or pale in colour.

2 You need soap flakes, a liquid soapless detergent and a solid soapless detergent. Solutions in tap water of 1% concentration will be suitable.

3 This is a long project. You may like to split the work up and have different people or groups tackle different fabrics.

4 Plan a series of tests on washing pieces of cloth in the different detergents in warm water.

5 Plan a series of tests on washing pieces of cloth in the different detergents in hot water.

6 Plan control experiments.

7 Decide how long the washing time will be in all the tests.

8 Decide how you will rinse the cloth.

9 Decide how you will compare the results of the tests on a certain fabric.

10 When you have done your share of the work, compare notes with other groups. Recommend detergents and temperatures for washing the different fabrics.

Experiment 9.8
Matching the cleaner to the stain

Your project is to find out which stain-remover is the best for each kind of stain.

Cleaning substances
Soap solution, non-enzyme detergent, enzyme detergent, dry-cleaning fluid, borax solution (10% sodium borate), methylated spirit, white spirit.

Staining substances
Dust, crayon, biro, make-up, chocolate, chewing gum, shoe polish, gravy, blood, curry, egg, cream, ice-cream, tar, coffee, tea, milk, wine, beer, ink, felt-tip pen, paint, wax, scorch marks.

Experiment 9.9
A soil problem

Soil has to retain water in order to nourish the plants which grow in it. Which kind of soil is best at holding water: clay, sandy soil, chalky soil or loamy soil? Your problem is to find out.

1 Obtain samples of the types of soil listed.

2 Plan a set of tests which you could do to answer the question. Say how you can make a fair comparison. What factors will you keep the same for each soil tested? How will you add water to the soil samples? Sketch the apparatus that you will use. What measurements will you make?

3 When you have a plan of campaign, show it to your teacher. If your plan is approved, carry it out.

TOPIC 10 Metals

Experiment 10.1
Reactions of metals with water and with dilute acids

1 Set up a rack of test-tubes. Half fill each with water.

2 Add *one* of the following metals to each test-tube of water: zinc, calcium, magnesium, iron, tin, aluminium, copper, lead.

3 If there is no reaction with cold water, warm the test-tube, and then leave it to stand.

4 Work through the list of metals, one at a time. Tabulate your results (Table A).

Table A Reactions of metals with water

Metal	*Reaction with cold water*	*Reaction with hot water*
Aluminium		
Calcium		
Copper		
Iron		
Lead		
Magnesium		
Tin		
Zinc		

5 There are two metals to add to your list. Sodium and potassium react violently with water. Perhaps your teacher will demonstrate their reactions.

6 Revise the action of dilute acids on metals. Take a rack of test-tubes, and half fill them with dilute hydrochloric acid. Add a different metal to each test-tube. See whether there is a reaction. If there is no reaction, warm the test-tube. Look for signs of reaction.

7 Tabulate your results (Table B).

Table B Reactions of metals with dilute acid

Metal	*Reaction with cold hydrochloric acid*	*Reaction with hot hydrochloric acid*
Iron		
etc.		

Extension

How fast is the reaction between a metal and an acid? It depends, as you will have seen, on which metal you mean and which acid you mean. Can you design an apparatus which will enable you to collect the gas given off? Then you can time how long it takes for a certain volume of gas to collect.

How long does it take for, say, 10 cm^3 of gas to collect when zinc reacts with

a hydrochloric acid,
b sulphuric acid,
c ethanoic acid?

What do the results tell you about these acids?

How long does it take for 10 cm^3 of gas to collect when you allow hydrochloric acid to react with a piece of

a zinc,
b iron,
c magnesium,
d any other metal you can test?

What do your results tell you about the reactivity of these metals? Are there any other factors which might affect the rate of the reaction?

Experiment 10.2
Studying the reaction of calcium with water

1 Fill a 250 cm^3 beaker three-quarters full with water. Drop in a piece of calcium.

2 Fill a test-tube with water, and invert it over the calcium (see Fig. 10.2a). Collect the gas formed. When the test-tube is full, test the gas.

3 Put your thumb over the end of the test-tube which is under water (see Fig. 10.2b). Take the tube out of the water, and open it at a flame (see Fig. 10.2c). What happens? What gas gives this test?

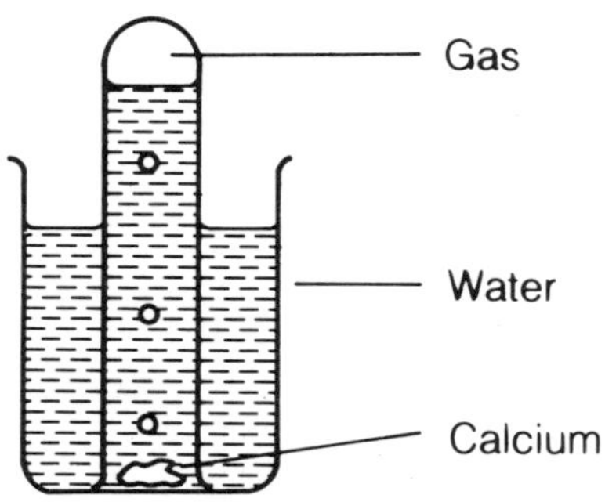

Fig. 10.2(a) Calcium and water

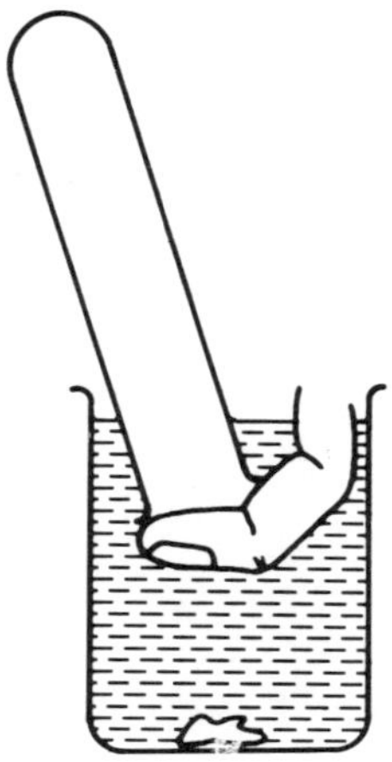
Fig. 10.2(b) Closing the test-tube

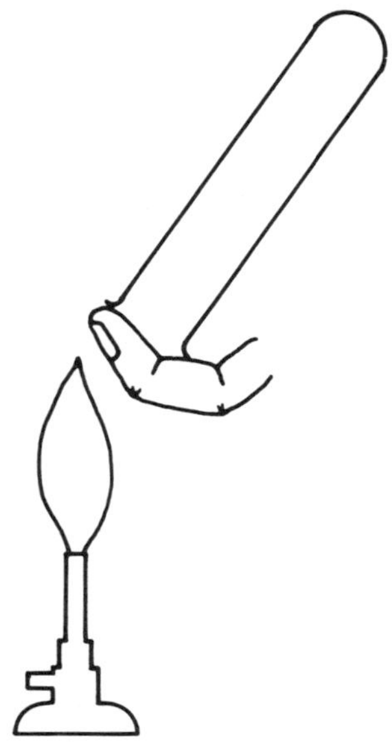
Fig. 10.2(c) Opening the test-tube

4 Filter the contents of the beaker.

5 Pour 10 cm^3 of the filtrate into a test-tube. Add 3 drops of litmus solution. Note the result, and say what this test tells you.

6 Pour 10 cm^3 of the filtrate into a second test-tube. Blow into it through a straw. Note the result. What solution gives this reaction?

7 Now put all your observations together.

calcium + water → **A**(g) + **B**(aq)

The test in step **3** shows that the gas **A** is ———.
The tests in steps **5** and **6** show that the solution is ———.
The word equation is

calcium + water →

The chemical equation is

Ca(s) + ——— →

Experiment 10.3
Studying the reaction of magnesium with steam

You will have found from Experiment 10.1 that magnesium reacts slowly with cold water. In this experiment, you can find out what happens when you increase the temperature.

1 Take a small wad of rocksil. Let it soak up as much water as possible. Then pack the wet rocksil into a Pyrex test-tube to a depth of 3 cm.

2 Clamp the test-tube in a horizontal position (see Fig. 10.3). Put magnesium powder into the middle of the test-tube. Fit a delivery tube.

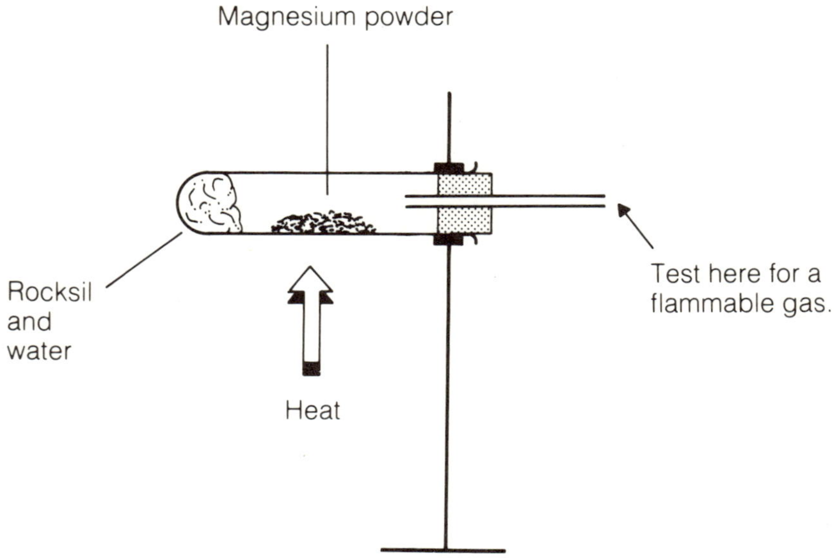

Fig. 10.3 Magnesium and steam

3 **Wear safety glasses**. Heat the magnesium. Enough heat will reach the water to turn it into steam.

4 From time to time, move the Bunsen flame to the end of the delivery tube. Is there a flammable gas given off?

5 When the reaction is over, tip out the contents of the test-tube. Add water, and warm. Divide the solution into two parts. Add a few drops of blue litmus to one part of the solution. Add a few drops of red litmus to the other part.

6 Collect your observations together. Copy and complete the following passage.

Magnesium + steam → gas + solid

When the gas met a Bunsen flame, it ————.
The solid dissolved in cold water/hot water to give a solution.
The solution turned blue litmus ————.
The solution turned red litmus ————.
The gas is likely to be ————.
The solid is likely to be ————.
The word equation for the reaction is

Experiment 10.4
Studying the action of steam on iron

1 Take a small wad of rocksil, and soak up as much water as it will hold. Push it into a Pyrex test-tube until you have a 3 cm length of rocksil.

2 Push a wad of iron wool into the test-tube. Attach a delivery tube. Arrange to collect any gas that is formed, as shown in Fig. 10.4a. See that the test-tube ready to receive gas is full of water.

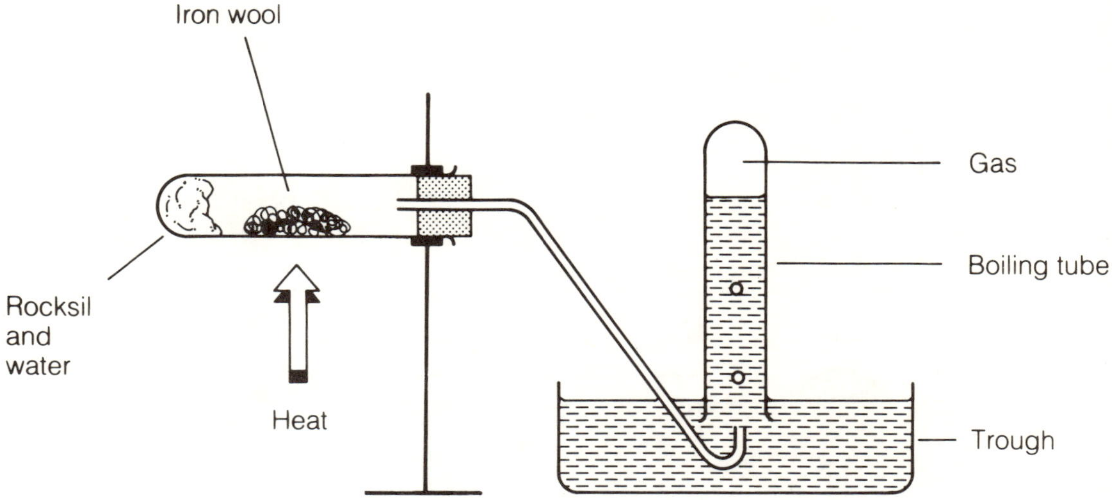

Fig. 10.4(a) Heating iron in steam

3 Heat the part of the test-tube containing iron wool. Do not heat the rocksil directly. Move the flame to and fro so as to keep the metal hot and also to vaporise the water in the rocksil. You need to keep a steady flow of steam passing over the hot metal.

4 Collect any gas formed. When the test-tube is full of gas, put your thumb over the open end. Remove it from the trough. Open it at a flame. What happens?

5 As soon as you stop heating, lift the delivery tube out of the trough. This precaution will avoid 'sucking back' (water passing up the delivery tube when the pressure of gas in the heated tube drops). You do not want cold water to reach the hot tube and crack it.

6 When the Pyrex tube cools, tip out the iron wool, and examine it. What changes do you see?

7 Bring your observations together. Copy and complete the following passage.

Iron + steam → gas + solid

The gas was tested by ————. This showed it to be ————.
The solid on the surface of the iron wool appears ————.
In fact, this solid is an oxide of iron with the formula Fe_3O_4.
A word equation for the reaction is

iron + steam → ———— + ————

A balanced chemical equation is

$Fe(s) + H_2O(g) \rightarrow$ ———— + ————

8 Repeat the experiment using zinc instead of iron.

9 You have now studied the reaction of a number of metals with water and steam. Place the metals in rank order, with the most reactive first.

Extension

When iron and zinc react with steam, the reaction is

metal + steam → metal oxide + hydrogen

The idea of this extension is to see whether there are metals which will give the reverse reaction

metal oxide + hydrogen → metal + steam

1 Set up the apparatus shown in Fig. 10.4b.

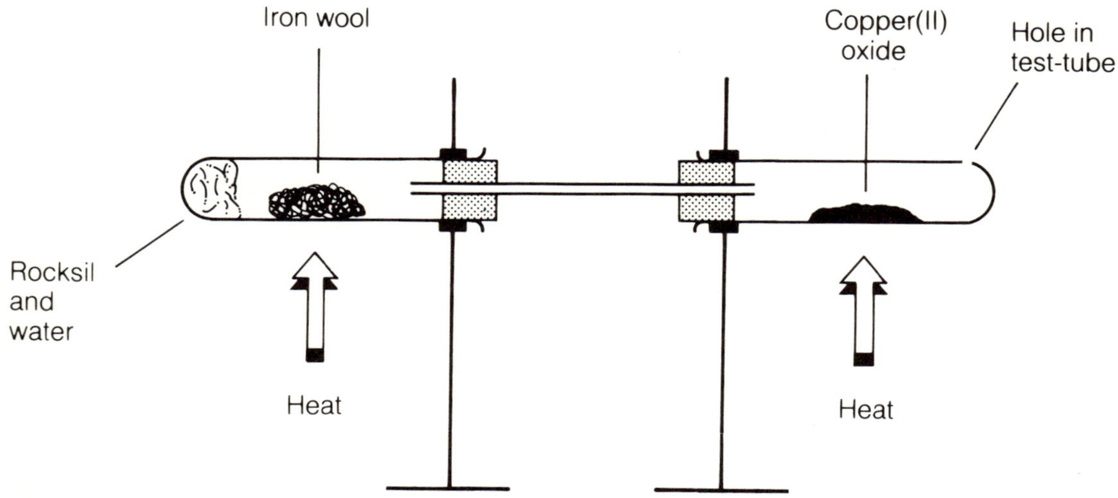

Fig. 10.4(b) Passing hydrogen over heated copper(II) oxide

2 Light the Bunsens under the iron wool and the copper(II) oxide. As you heat the iron wool, hydrogen is formed in the left-hand side of the apparatus.

3 Note the appearance of the heated copper(II) oxide. Hold a light to the hole in the right-hand test-tube. If hydrogen is leaving the tube, it should light.

4 Make a note of the changes which you observe. Deduce what reaction has occurred. Copy and complete the following passage.

In this experiment, hydrogen was made by the action of ———— on ————.

When hydrogen passes over heated copper(II) oxide, we see ————.

The reaction that has occurred is

hydrogen + copper(II) oxide → ———— + ————

$H_2(g)$ + $CuO(s)$ → ———— + ————

Hydrogen has taken ———— from copper(II) oxide. Copper(II) oxide has been ———— by hydrogen. The chemical name for this type of reaction is ————.

Experiment 10.5
Putting metals in order of reactivity

You have seen from the experiments you have done with metals and water and with metals and dilute acids that some metals are more reactive than others. You have been able to put some metals into a rank order of reactivity. Another way of ranking metals in order of reactivity is to study *displacement reactions*. You have found from your experiments that magnesium is more reactive than copper. Will magnesium displace copper from copper salts?

1 Take a test-tube of copper(II) sulphate solution. Drop a piece of magnesium ribbon into it. Observe what happens.

2 What change do you see in the copper(II) sulphate solution?
What change do you see in the piece of magnesium?
Now you should be able to answer the question, 'does magnesium displace copper from copper salts?'

3 Set up the experiments shown in Fig. 10.5a. Leave the solutions to stand. Record any changes you see.

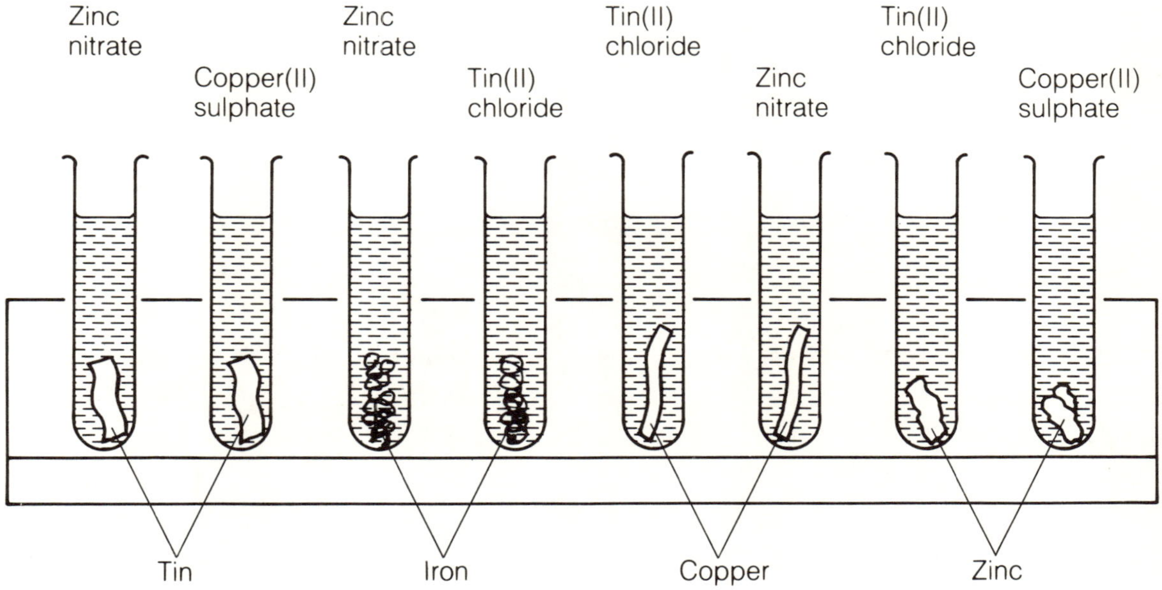

Fig. 10.5(a) Displacement reactions

4 Copy and complete the following word equations. If no reaction occurs, write 'No reaction'.

magnesium + copper(II) sulphate →

tin + zinc nitrate →

tin + copper(II) sulphate →

iron + zinc nitrate →

iron + tin(II) chloride →

copper + tin(II) chloride →

copper + zinc nitrate →

zinc + tin(II) chloride →

zinc + copper(II) sulphate →

5 Say what experiment you could do to tell you which is the more reactive, copper or lead. Do the experiment you have planned.

6 From your results, say which metal is the most easily displaced from its salts. This is the least reactive of the metals.

Which metal is the most difficult to displace from its salts? This is the most reactive of the metals you have studied.

Arrange the metals in order of decreasing reactivity.

Extension 1

Perhaps you would like to grow some crystals of lead. Set up the apparatus show in Fig. 10.5b. Leave it to stand for a day. Describe what you see. Copy and complete the word equation for the reaction:

zinc + lead(II) ethanoate →

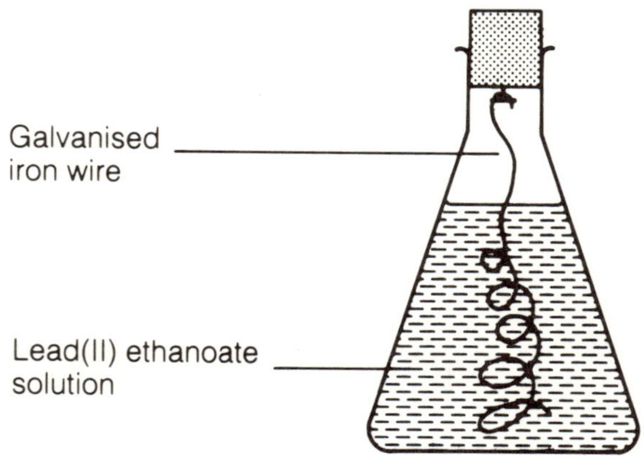

Fig. 10.5(b) Growing lead crystals

Extension 2

You can grow silver crystals by setting up the apparatus shown in Fig. 10.5c. Describe what you see. Copy and complete the word equation:

copper + silver nitrate →

When your crystals have grown, let your teacher have them for recycling.

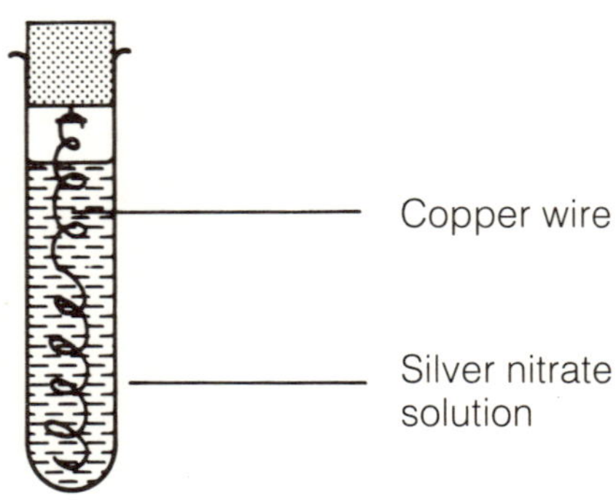

Fig. 10.5(c) Growing silver crystals

Experiment 10.6
Cleaning the silver — an experiment to do at home

1 Dissolve 30 g (one tablespoonful) of sodium carbonate crystals (washing soda) in 200 cm^3 (one cupful) of warm water.

2 Into the solution put a tarnished silver article. Add some aluminium foil (baking foil or clean milk bottle tops) (Fig. 10.6).

3 From time to time, inspect the silver. When it is clean, rinse it with water.

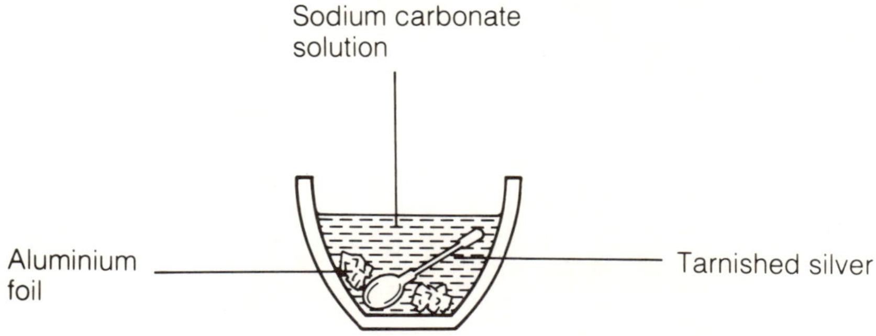

Fig. 10.6 Cleaning the silver

4 Can you explain what happens? Remember
  a aluminium is a more reactive metal than silver,
  b tarnish is a silver compound, silver sulphide.

5 Paula believes that, when silver polish removes the layer of tarnish from the surface of a silver object, it gradually wears away the surface of the silver object. After your experiment, you should now understand that silver polishes do not work in this way. Write a few sentences to explain to Paula why it is safe to polish the silver. Tell her what type of chemical reaction is involved.

Experiment 10.7
What happens when you heat carbon with a metal oxide?

1 Assemble a number of metal oxides: aluminium oxide, calcium oxide, copper(II) oxide, iron(III) oxide, lead(II) oxide, zinc oxide and any others you wish to use.

2 Mix aluminium oxide with powdered wood charcoal (carbon) on a piece of ceramic paper.

3 Heat the mixture on a tripod and gauze (see Fig. 10.7). Observe any changes that occur.

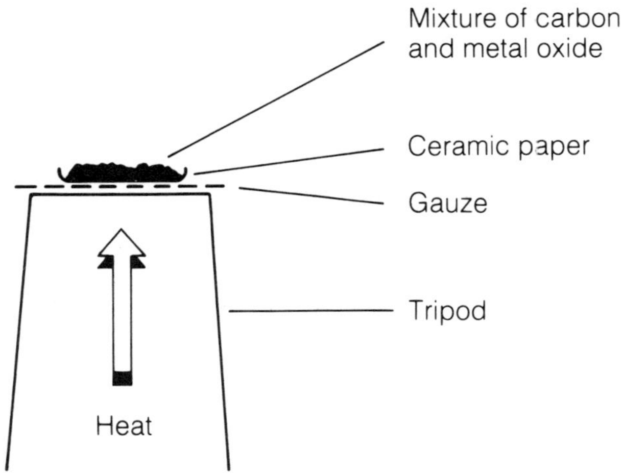

Fig. 10.7 Heating a metal oxide with carbon

4 Repeat steps **2** and **3** with the other metal oxides, one at a time.

5 In some cases, after heating, you will see a bead of material which looks metallic. You can test it to find out whether it conducts electricity. Construct a test circuit, and use it to find out whether your product is a metal.

6 Copy and complete this passage

The removal of oxygen from a substance is called ————. Carbon is a ———— agent. The following list of metals can be obtained by ———— their oxides with carbon:

————————————————————————————————

The metals whose oxides are easily ———— by carbon are more/less reactive than the metals whose oxides do not react with carbon.

Experiment 10.8
Electricity from metals

Zinc is fairly high in the reactivity series of metals. Copper is low in the reactivity series.

1 Set up the apparatus shown in Fig. 10.8. G is a galvanometer or a multimeter.

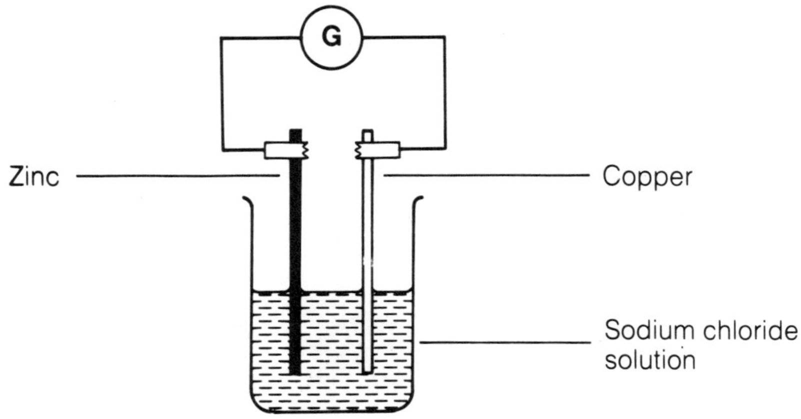

Fig. 10.8 A combination of two metals

2 Record the electromotive force (e.m.f.) in volts measured by the galvanometer.

3 Iron and zinc are closer together in the reactivity series than copper and zinc. If you connect them as shown in Fig. 10.8, will the e.m.f. be the same as in step **2** or greater or less than in step **2**? Try it, and see whether you are right.

4 Connect a strip of iron with a strip of copper as shown in Fig. 10.8. Record the e.m.f. Compare this e.m.f. with the e.m.f. in step **1**. What does this comparison tell you about the positions of iron and zinc in the reactivity series?

5 You are given a strip of an unknown metal. Describe how you could find out where the metal is in the reactivity series compared with zinc, iron and copper. Consult your teacher to see whether there is a mystery metal for you to examine.

Experiment 10.9
Focus on iron

The characteristics of a piece of iron or steel depend on the heat treatment that it has received.

1 Examine a nail. Bend it with pliers. Notice how easy or difficult it is to bend.

2 Heat a nail in a Bunsen burner until it is red hot (800 °C). Slowly lift it out of the flame over a 1 minute period. Put it on a heat-proof surface to cool slowly to room temperature. You have *annealed* it. When it is cold, try bending it with pliers. Is the metal more or less pliable than the nail in step **1**?

3 Have ready a beaker of cold water. Heat the third nail to red heat (800 °C). Then take it out of the flame, and plunge it into the cold water. You have *quenched* it.

4 With the fourth nail, repeat step **3**. Then rub the surface of the nail with emery paper. Heat it for 5 minutes just above a Bunsen burner flame (at 300 °C). Do not let the metal become red hot. When it turns bright blue, take it out of the flame. Place the nail on a heat-proof surface to cool. You have *tempered* it.

5 Test all the nails for hardness by scratching them with a file.
Test how pliable all the nails are by bending them with pliers.

6 Tabulate your results.

Table Results of heat treatment

Nail	*Process*	*Results of scratching and bending tests*
1		
2		
3		
4		

What kind of treatment would you use on iron which was to be used for
a a hammer
b a sword,
c a wheel?

Experiment 10.10
Investigating rusting

When iron rusts, it is converted into hydrated iron(III) hydroxide, $Fe(OH)_3 \cdot nH_2O$. The first step in the process is

iron atoms → iron(II) ions + electrons

$$Fe(s) \rightarrow Fe^{2+}(aq) + 2e^-$$

When the electrons released react with oxygen and water, hydroxide ions, OH^-, are produced. Eventually, iron(III) hydroxide forms. It is a rusty-red solid.

Rusting is an expensive problem. It is important to be able to carry out experiments on methods of treating iron to prevent or slow down rusting. Normally, such experiments take a long time. There is, however, an indicator called *ferroxyl* which enables you to do rusting experiments more quickly. Ferroxyl reacts with iron(III) ions to give a blue colour. In this way, it can detect the first signs of rusting. Ferroxyl also reacts with hydroxide ions to give a pink colour.

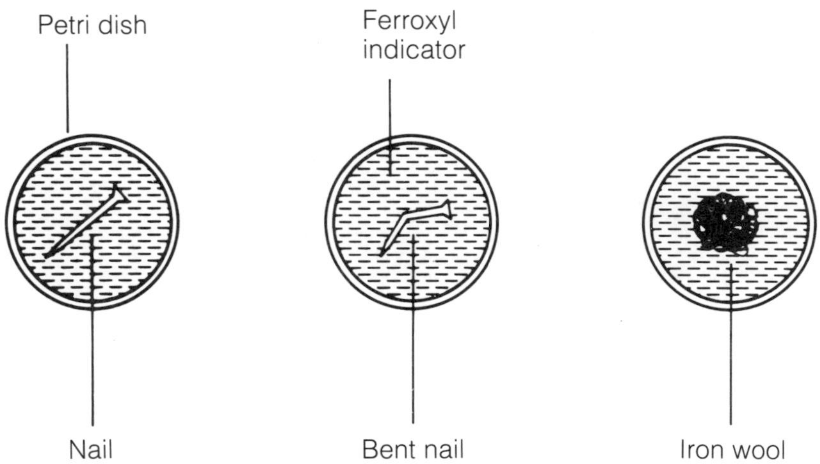

Fig. 10.10 Ferroxyl indicator experiments

1 Fill a petri dish with ferroxyl indicator. Place a clean iron nail in the dish (see Fig. 10.10). Watch carefully and note any colour changes. Where do the first signs of rusting occur?

2 With a pair of pliers, bend a clean iron nail to and fro twice in the same place. Put the nail into a petri dish of ferroxyl indicator. Note carefully where the colour changes occur. Has being bent affected the nail?

3 Roughen the surface of a clean iron nail with a file. Drop the nail into a petri dish of ferroxyl indicator. Note what happens.

4 Take a small wad of iron wool, and immerse it in a dish of indicator.

5 Which rusts the faster, an iron nail or a wad of iron wool?
Which rusts the faster, a nail you have bent or an untreated nail?
Which parts of a nail rust fastest?

6 You studied the effect of various heat treatments on the flexibility of iron nails in Experiment 10.9. Now you can find out how these heat treatments affect the speed with which iron rusts.

7 Holding a nail in tongs, heat it strongly until it is red hot. Put it on to a heat-proof mat to cool. You have *annealed* this nail.

8 Heat a second nail to red heat. Plunge it into cold water. You have *quenched* this nail.

9 Take a third nail, and repeat step **8**. Then heat the nail to blue heat and let it cool. You have *tempered* this nail.

10 Fill four petri dishes with ferroxyl indicator. Label them *annealed*, *quenched*, *tempered* and *control*. Put the three nails you have heated in steps **7**, **8** and **9** into three of the dishes and an untreated nail into the fourth dish.

Extension

How can you find out the effect on the speed with which iron rusts of
a a coat of paint,
b a layer of grease,
c galvanising (zinc-plating)?
Plan an experiment you could do, check your plan with your teacher, and then carry out your experiment.

Experiment 10.11
Corrosion of pairs of metals

This is another experiment which uses ferroxyl indicator. Remember: this indicator turns blue in the presence of iron(II) ions and red in the presence of hydroxide ions. In this experiment, the indicator is made into a gel with agar. The gel has the advantage that the colours that develop spread out less than they do in aqueous solution.

1 Pour 100 cm^3 of ferroxyl indicator solution into a beaker. Add 2 g of agar. Heat the mixture, with stirring, until it boils. Fill four specimen tubes with the hot solution.

2 With pliers, fasten a piece of magnesium ribbon round a clean iron nail. Treat other nails with aluminium foil, zinc foil, tin foil or copper foil.

3 Wait until the agar solution cools and starts to gel. Then drop a nail into each specimen tube (Fig. 10.11a).

4 Record the time. Watch closely. Note where the blue colour appears. Note where the red colour appears. Note any gas evolution. Note the time that has elapsed since you put the nail into the indicator.

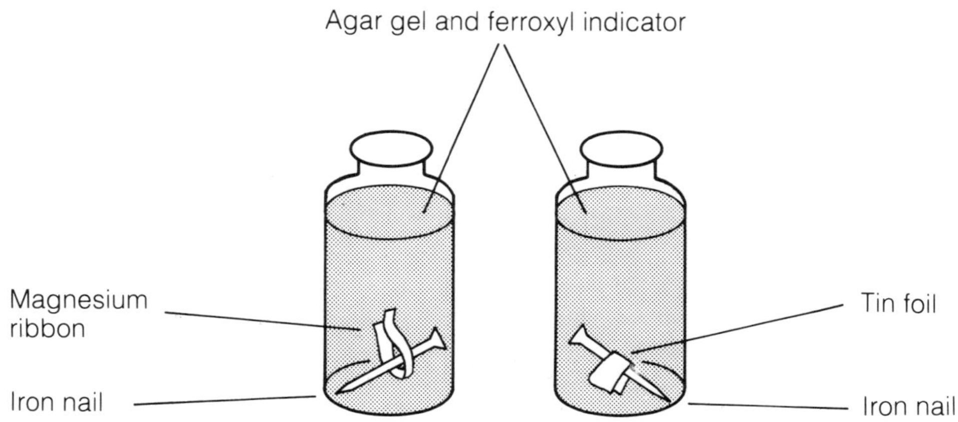

Fig. 10.11(a) Corrosion of pairs of metals

5 Tabulate your results.

Table Colours of ferroxyl indicator in corrosion tests

Metal	*Colour at metal*	*Colour at nail*	*Time before colours develop*

6 Say which of the metals that you tested make iron rust faster.
Where do these metals occur in the reactivity series of metals?
Say which of the metals that you tested slow down the rusting of iron.
Where do these metals occur in the reactivity series of metals?
What practical use is made of the metals which protect iron from rusting?

Extension

Is it necessary for the second metal to be fastened tightly to iron in order to affect rusting? Try connecting a second metal to iron with a wire. Set up the apparatus shown in Fig. 10.11b. Which do you predict will rust faster, nail A or nail B? Try the experiment and see whether you are right.

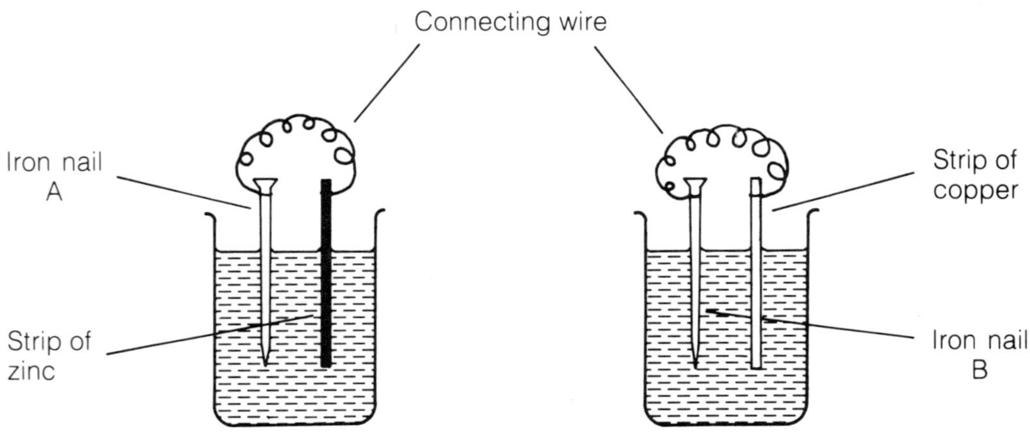

Fig. 10.11(b) Nails in agar gel containing ferroxyl indicator

Experiment 10.12
Focus on aluminium

Mercury salts are poisonous. Do not handle mercury(II) chloride with your fingers.

1 Take a piece of aluminium foil. Hold a wad of cotton wool in tongs. Dip it into a solution of mercury(II) chloride. Rub a circular patch on the aluminium foil with the cotton wool (see Fig. 10.12).

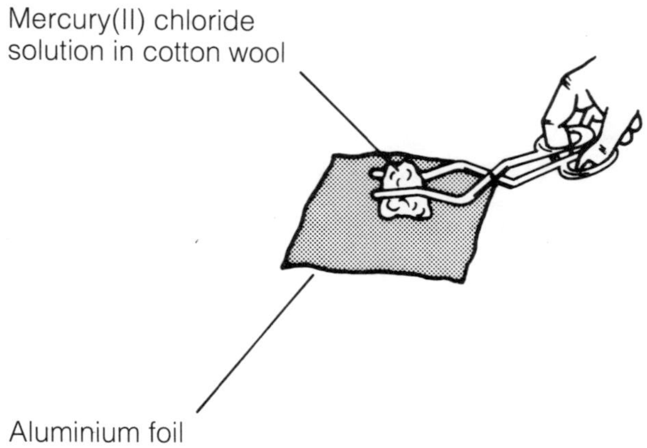

Fig. 10.12 Rubbing aluminium foil

2 Leave the foil to stand for half an hour. Notice what happens.

3 Take a second piece of aluminium foil. Rub it with cotton wool soaked in mercury(II) chloride as before. After ½ minute, wash off the mercury(II) chloride. Drop the foil into a beaker of warm water. Notice what happens.

4 Immerse a piece of aluminium foil (10 cm square) in a solution of mercury(II) chloride for ½ minute. Take the foil out of the solution, and wrap it round the bulb of a thermometer. What evidence have you that a chemical reaction is happening?

5 Aluminium is unreactive because it is coated with a layer of aluminium oxide. Any chemical which can remove this layer enables aluminium to show its true reactivity. Can you work out how mercury(II) chloride does this? Remember that mercury is able to dissolve metals to form *amalgams*.

Experiment 10.13
Removing the oxide film from aluminium

Aluminium is unreactive because it is coated with a layer of aluminium oxide. This experiment is about removing the oxide layer.

1 Line a petri dish with aluminium foil. Pour in copper(II) sulphate solution. Is there a reaction?

2 Add a little sodium chloride. Is there a reaction?

3 Take a piece of aluminium foil, and add sodium chloride. Is there a reaction?

4 From your observations, you have found that the oxide film is removed when four ions are present. Write down the names of these four ions.

5 You can do an investigation to find out whether *all* these ions are needed to remove the oxide film. First make a copy of Table A.

6 Take a rack containing four test-tubes. Put a piece of aluminium foil into each one. Add 5 cm^3 of copper(II) sulphate solution to each test-tube. Fill them as described in Table A.

Table A Aluminium and solutions of salts

Test-tube	*Chemicals present with aluminium*	*Ions present*	*Result*
1	Copper(II) sulphate solution		
2	Copper(II) sulphate solution, sodium chloride		
3	Copper(II) sulphate solution, sodium sulphate		
4	Copper(II) sulphate solution, potassium chloride		

7 Write into your copy of Table A a list of the ions present in each solution. Record whether or not there was a reaction. From your results so far, answer these questions. For the removal of the oxide layer from aluminium, do you need to have present

a sodium ions,
b potassium ions,
c chloride ions?

You have put copper(II) sulphate into all four test-tubes. You will have to do further experiments to find out whether copper ions and sulphate ions are essential for the reaction.

8 Make up a further rack of four test-tubes. Put a piece of aluminium into each one. Fill the test-tubes as shown in Table B. Use 5 cm^3 of each solution.

Table B

Test-tube	*Chemicals present with aluminium*	*Ions present*	*Result*
1	Sulphuric acid, sodium chloride		
2	Sulphuric acid, potassium chloride		
3	Hydrochloric acid, copper(II) sulphate		
4	Hydrochloric acid, copper(II) nitrate		

9 From your results, answer the following question. For the removal of the oxide layer, do you need the presence of

a sulphate ions,
b chloride ions,
c nitrate ions,
d copper(II) ions?

10 There are two ions which are present in all the experiments in which aluminium loses its oxide layer and shows its true reactivity. What are the two ions?

11 What compound possesses these two ions? Let us call this compound X. Test to see whether you are right in thinking that X will make aluminium react. Do the set of test-tube experiments outlined in Table C.

Table C Tests on compound X

Test on X	*Add to a solution of X*	*Result*
Test 1	Aluminium foil	
Test 2	Aluminium + hydrochloric acid	
Test 3	Aluminium + sulphuric acid	

TOPIC 11 Formulas and equations

Experiment 11.1
Finding the formula of magnesium oxide

The idea of this experiment is to burn a known mass of magnesium in air and to find the mass of magnesium oxide produced.

1 Weigh a crucible with its lid.

2 Clean a 10 cm length of magnesium ribbon. As it has been machined, the ribbon is oily. Remove the oil with a solvent, e.g. dichloromethane.

3 Coil the magnesium ribbon. Put it into the crucible. Weigh the crucible with its lid and contents.

4 Place the crucible on a pipeclay triangle supported on a tripod (see Fig. 11.1). Heat the crucible. From time to time, lift the lid with tongs. This will let in air. Do not lift the lid too far; you do not want any of the magnesium oxide that is formed to blow away. Continue heating until all the magnesium has burned.

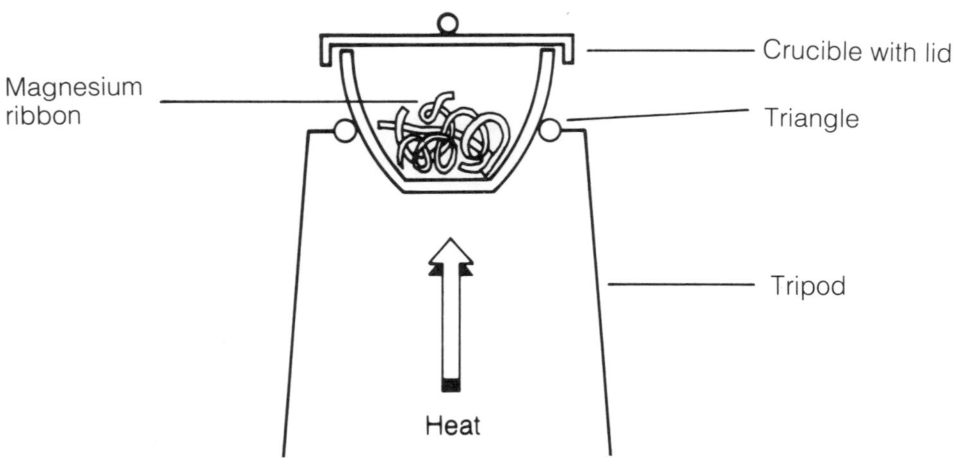

Fig. 11.1 Heating magnesium ribbon

5 Let the crucible cool. Then reweigh the crucible and lid and contents.

6 List your results.

Mass of crucible + lid = m_1 = ——— g.

Mass of crucible + lid + magnesium = m_2 = ——— g.

Mass of crucible + lid + magnesium oxide = m_3 = ——— g.

Mass of magnesium = $m_2 - m_1$ = ——— g.

Mass of magnesium oxide = $m_3 - m_1$ = ——— g.

Mass of oxygen combined with magnesium = $m_3 - m_2$ = ——— g.

The relative atomic masses are A_r(Mg) = 24 and A_r(O) = 16.

The empirical formula of magnesium oxide is ———.

Experiment 11.2
Finding the formula of copper oxide

The idea in this experiment is to reduce a known mass of copper oxide to copper. A suitable apparatus is shown in Fig. 11.2a.

1 Weigh a Pyrex test-tube with a hole in the end.

2 Put two spatula measures of copper oxide into the middle of the test-tube. Weigh the test-tube + copper oxide.

3 Assemble the apparatus shown in Fig. 11.2a. Turn on the gas. Let gas pass through the apparatus for ½ minute to remove air. Then light the gas which escapes from the hole in the test-tube. Turn down the gas until the flame is about 5 cm high.

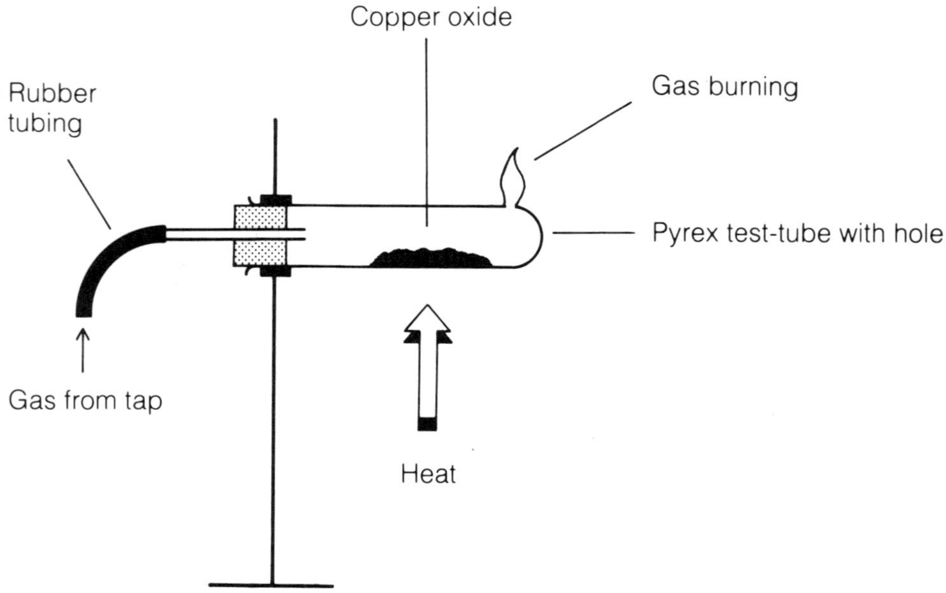

Fig. 11.2(a) Reducing copper oxide

4 Light the Bunsen burner under the copper oxide. Heat until no further change seems to be happening.

5 Turn off the Bunsen. Leave gas flowing through the apparatus while the apparatus is cooling.

6 Switch off the gas. Weigh the test-tube and contents.

7 List your results.

Mass of test-tube = m_1 = ———— g.

Mass of test-tube + copper oxide = m_2 = ———— g.

Mass of test-tube + copper = m_3 = ———— g.

Mass of copper oxide = $m_2 - m_1$ = ———— g.

Mass of copper = $m_3 - m_1$ = ———— g.

Mass of oxygen = $m_2 - m_3$ = ———— g.

The relative atomic masses are $A_r(Cu) = 63.5$ and $A_r(O) = 16$.

Amount in moles of copper = $(m_3 - m_1)/63.5$ = ——— mol.

Amount in moles of oxygen = $(m_2 - m_3)/16$ = ——— mol.

Ratio: amount of copper/amount of oxygen = ——— .

Empirical formula for copper oxide is ——— .

Alternative apparatus

This experiment works better with the copper oxide in a porcelain boat. If you have enough porcelain boats to go round the class, please use them. Put the porcelain boat inside the Pyrex test-tube with a hole (see Fig. 11.2b). Adapt the instructions; when the instructions say weigh the test-tube, you weigh the porcelain boat. You should get good results!

Another improvement is to put a small piece of *meta fuel* into the boiling tube near the bung, as shown in Fig. 11.2b. The vapour of *meta fuel* will speed up the reduction of copper oxide to copper.

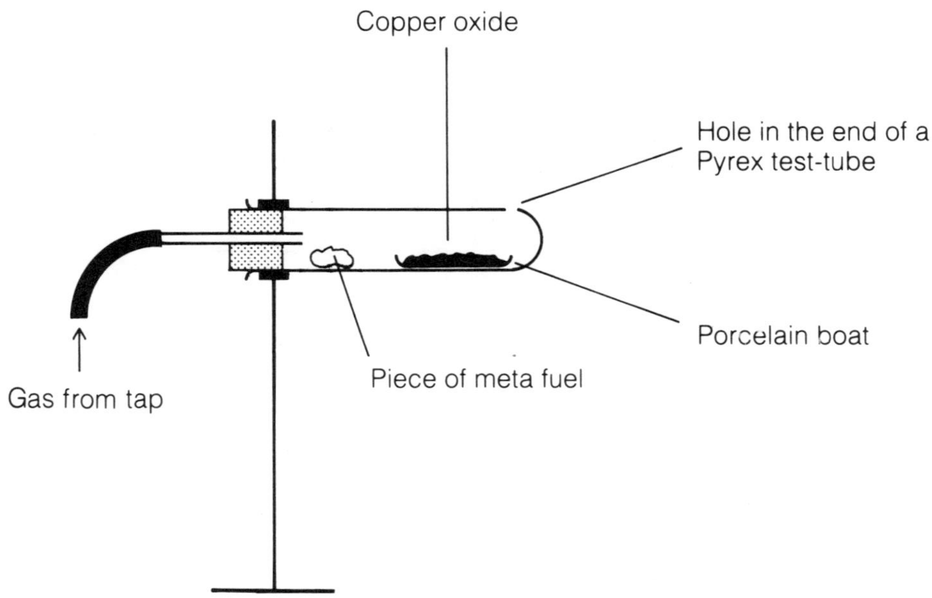

Fig. 11.2(b) Alternative apparatus

Experiment 11.3
Finding the balanced equation for a reaction

The reaction you are to study is the reaction between solutions of lead(II) nitrate and potassium iodide to give a precipitate of lead(II) iodide.

You need a solution of potassium iodide of concentration 0.5 mol/l (83 g/l) and a solution of lead(II) nitrate of concentration 0.5 mol/l (165.5 g/l).

1 Put six test-tubes into a rack.

2 Fill one burette with potassium iodide solution and a second with lead(II) nitrate solution.

3 Run 4.0 cm^3 of potassium iodide solution into each test-tube.

4 Run a different volume of lead(II) nitrate solution into each test-tube, as shown in the table.

5 Make a copy of the table.

Table Solutions to mix

Test-tube	*Volume in cm^3 of 0.50 mol/l potassium iodide*	*Volume in cm^3 of 0.50 mol/l lead(II) nitrate*	*Amount in mol of potassium iodide*	*Amount in mol of lead(II) nitrate*	*Height in mm of precipitate* **a**	**b**
1	4.0	0.50	2.0×10^{-3}	2.5×10^{-4}		
2	4.0	1.0				
3	4.0	1.5				
4	4.0	2.0				
5	4.0	3.0				
6	4.0	4.0				

6 Shake each test-tube to mix the solutions thoroughly.

7 Allow the precipitates to settle. Then measure the height in mm of the precipitate of lead(II) iodide in each test-tube. Enter heights in the table.

8 After ¼ hour, measure the heights again, and enter them.

9 Calculate and fill in the amounts (in moles) of potassium iodide and lead(II) nitrate in each test-tube.

10 On graph paper, plot the height in mm of the precipitate (column **a**) against the volume in cm^2 of lead(II) nitrate used. Connect the points. Can you explain the shape of the graph? Repeat for column **b**.

11 From the graphs, select the combination in which there is just enough lead(II) nitrate solution to react with all the potassium iodide.

12 From the table, say what is the ratio
(amount in moles of lead(II) nitrate)/(amount in moles of potassium iodide)
for this combination.
Using this ratio, complete and balance the equation

$Pb(NO_3)_2(aq) + KI(aq) \rightarrow$

Experiment 11.4
Finding the formula for magnesium sulphate crystals

Magnesium sulphate crystallises with water of crystallisation as $MgSO_4 \cdot nH_2O$. By finding the mass of water lost when a known mass of crystals is heated at 100 °C, the formula can be found.

1 Weigh a clean dry crucible. Record its mass m_1.

2 Half fill the crucible with magnesium sulphate crystals and reweigh to give m_2.

3 Set up the apparatus shown in Fig. 11.4. Heat gently; the water of crystallisation will be driven off at 100 °C. When no more water vapour can be seen, turn off the Bunsen, and let the crucible cool.

4 Weigh the crucible to give m_3.

5 Reheat the crucible. Let it cool, and reweigh it to give m_4. If m_4 is lower than m_3, you will have to repeat steps **3** and **4**. You must heat to constant mass — until all the water of crystallisation has been driven off.

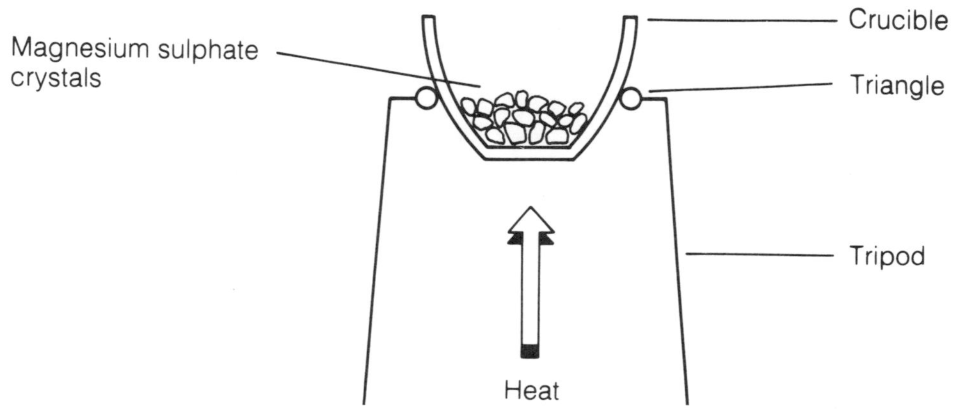

Fig. 11.4 Heating magnesium sulphate crystals

6 Calculate the formula of the crystals.

Mass of crucible = m_1

Mass of crucible + magnesium sulphate crystals = m_2

Mass of magnesium sulphate crystals = $m_2 - m_1$

Mass of crucible + anhydrous magnesium sulphate = m_4

Mass of anhydrous magnesium sulphate = $m_4 - m_1$

Mass of water of crystallisation = $m_2 - m_4$

Table Calculation of the formula of the crystals

Substance	Magnesium sulphate	Water
Formula	$MgSO_4$	H_2O
Mass	$m_4 - m_1$	$m_2 - m_4$
Molar mass	120	18
Amount in moles	$(m_4 - m_1)/120$	$(m_2 - m_4)/18$
Ratio of amounts	1 : ?	
Formula		

TOPIC 12 Carbonates and carbon dioxide

Experiment 12.1
Heating calcium carbonate

You can use a marble chip or a piece of limestone for this experiment. The idea is to find out whether there is a change in mass when calcium carbonate is heated.

1 Weigh a watch glass. Write down the mass.

2 Weigh a marble chip on the watch glass. Write down the mass.

3 Make a loop in the end of a piece of thick wire. Wedge the marble chip in this loop. Support the wire as shown in Fig. 12.1.

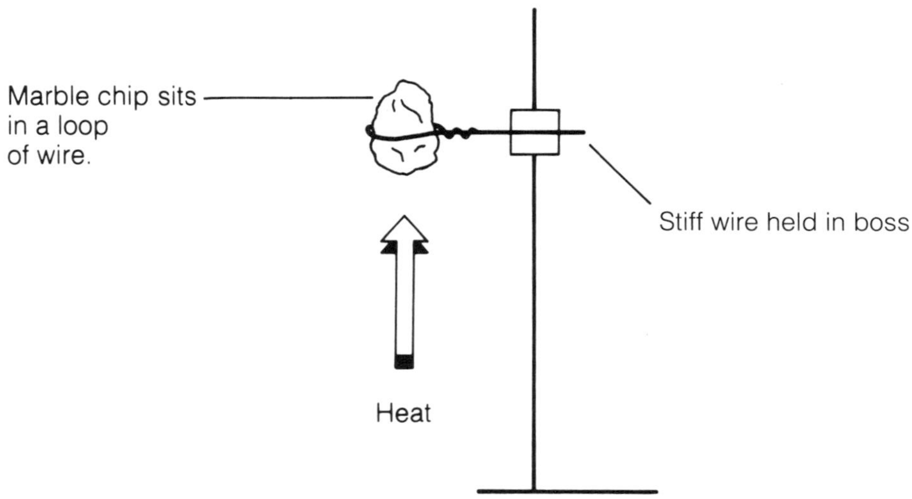

Fig. 12.1 Heating a marble chip

4 Heat the marble strongly for 15 minutes. Then turn off the Bunsen.

5 When the marble has cooled, tip it carefully on to the watch glass. Notice any change in the appearance of the marble. Reweigh the watch glass and marble chip. Write down the mass.

6 With a teat pipette, add water drop by drop to the marble chip. You should notice two things.

7 Add the mixture on the watch glass to 50 cm^3 of water in a beaker. Filter the solution. Divide the filtrate into two portions.

8 To one portion, add a few drops of universal indicator. Note the colour.

9 Blow into the second portion through a straw. What is the solution?

10 Now draw your observations together.

Mass of watch glass = ———— g

Mass of watch glass + marble chip = ———— g

Mass of watch glass + marble chip after heating = ———— g

Does the mass of the marble chip increase or decrease or stay the same?
What did you see when you added water to the chip after heating?
How can you explain this observation?
What colour did universal indicator turn? What does this prove?
What is the solution that is formed when the marble which has been heated reacts with water?
Copy this passage, filling in the names of **A**, **B**, **C** and **D**.

Calcium carbonate $\xrightarrow{\text{heated strongly}}$ **A**(s) + **B**(g)

A(s) + water → **C**(s)

C(s) + water → **C**(aq)

C(aq) + exhaled breath → **D**(s)

Write chemical equations for all the reactions.

Experiment 12.2
Making mortar

1 Take a plastic juice bottle, and cut off the top (Fig. 12.2a). You now have a disposable beaker.

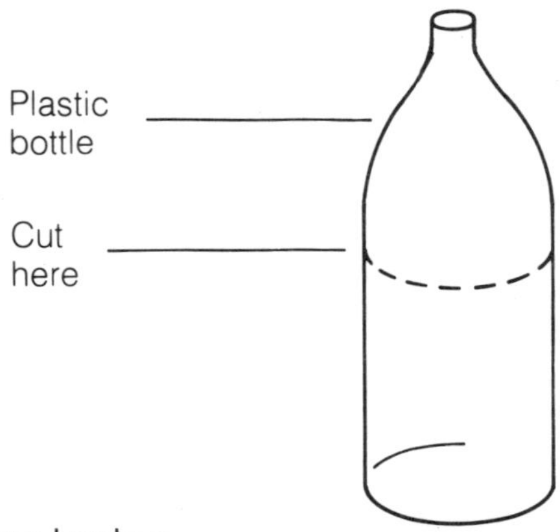

Fig. 12.2(a) Making a beaker

2 Tip equal volumes of sand, calcium oxide and water into the beaker. Mix thoroughly with a stick. You have made mortar.

3 Use some of the mortar to stick two bricks together.

4 Compare the mortar you have made with some mortar which has weathered for a year or more. Do the following test with both old and new mortar.

5 Assemble the apparatus shown in Fig. 12.2b. Add dilute hydrochloric acid to some new mortar.

6 Repeat step **5** with some weathered mortar.

7 What do you observe with **a** new mortar, **b** old mortar?

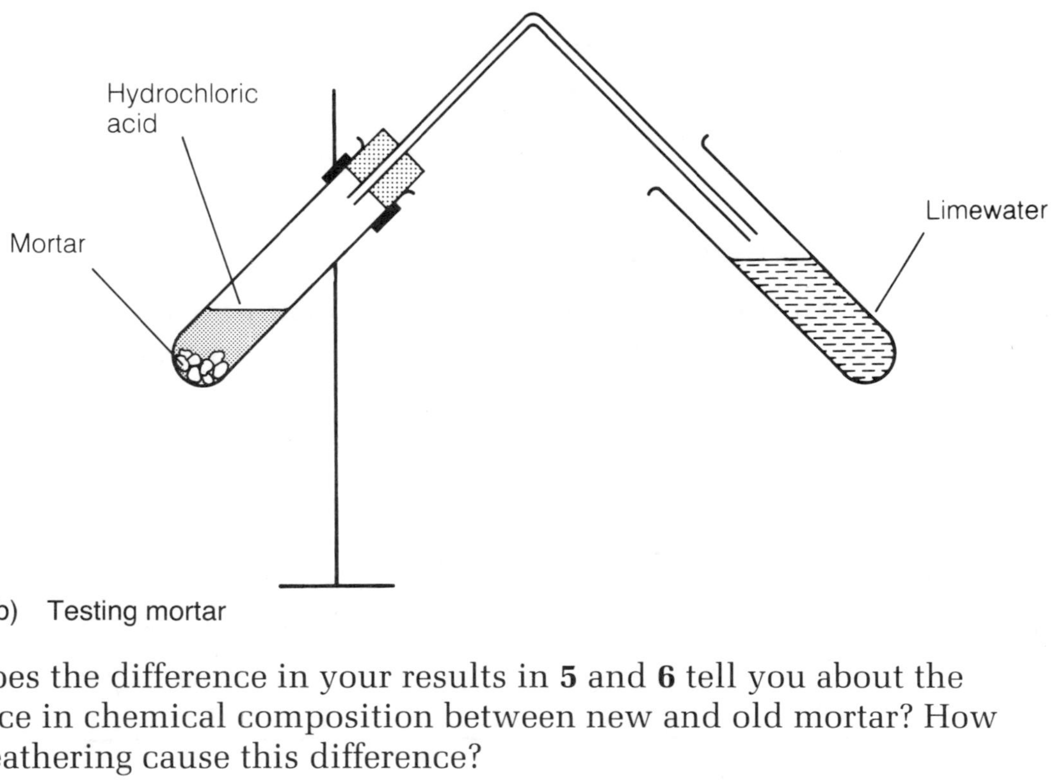

Fig. 12.2(b) Testing mortar

What does the difference in your results in **5** and **6** tell you about the difference in chemical composition between new and old mortar? How does weathering cause this difference?

Experiment 12.3
Investigating carbonates

Heating carbonates

Some carbonates give carbon dioxide on heating. You can test a number of carbonates by using the apparatus in Fig. 12.3a. Carbonates you can try are copper(II) carbonate, zinc carbonate, sodium carbonate, calcium carbonate and others. Tabulate your results.

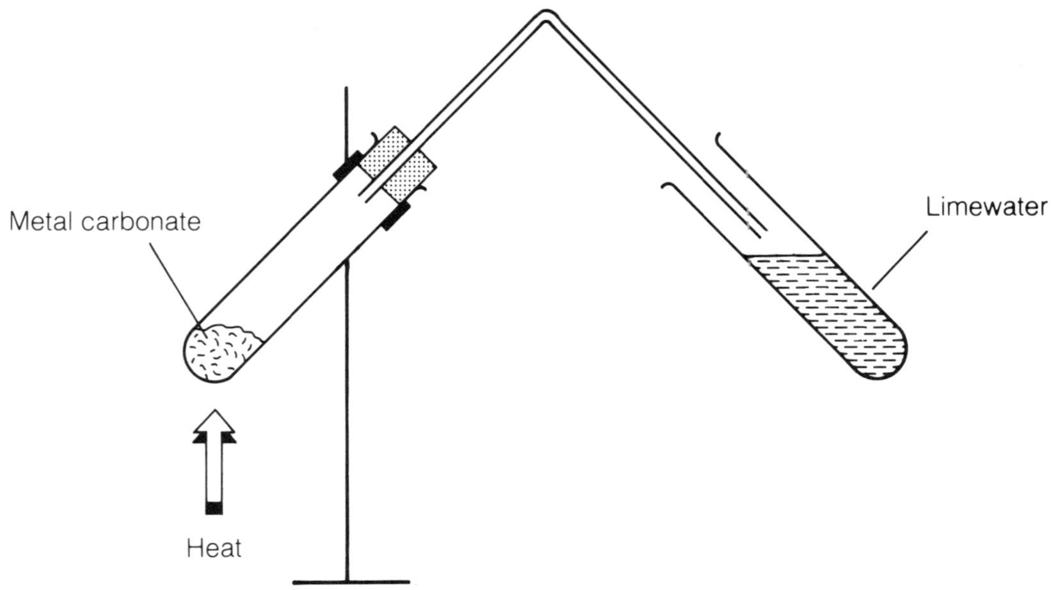

Fig. 12.3(a) Heating a carbonate

Table A Action of heat on carbonates

Carbonate	*Appearance*	*Observations on heating*
Copper(II) carbonate *etc.*		

Action of acids on carbonates

1 Can you obtain carbon dioxide from a carbonate and an acid? Figure 12.3b shows apparatus you can use to find out. Try calcium carbonate, sodium carbonate and others.

2 Tabulate your results.

Table B Action of hydrochloric acid on carbonates

Carbonate	*Appearance*	*Observations on adding acid*
Calcium carbonate etc.		

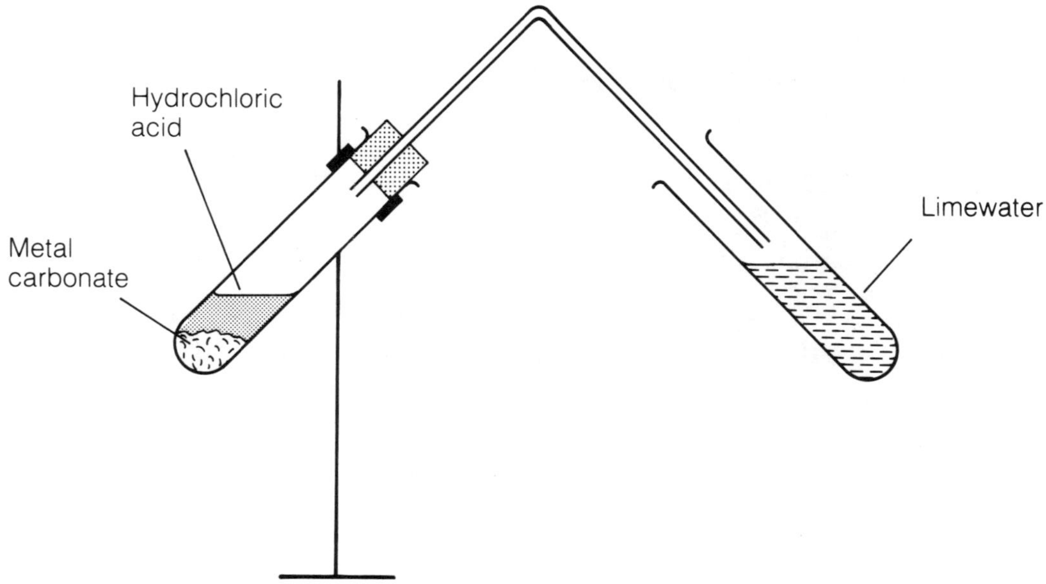

Fig. 12.3(b) Adding acid to a carbonate

3 Find out how dilute sulphuric acid reacts with sodium carbonate. Is there any difference between dilute sulphuric acid and dilute hydrochloric acid in this reaction?

4 Find out how dilute sulphuric acid reacts with calcium carbonate. Is there any difference between dilute sulphuric acid and dilute hydrochloric acid in this reaction?

5 Explain any differences you find in steps **4** and **5**.

Experiment 12.4
Preparation of carbon dioxide

Carbon dioxide is formed when an acid reacts with a carbonate. There are two ways of collecting the gas. As it is only slightly soluble in water, it can be collected over water. As it is denser than air, it can be collected downwards.

1 Set up the apparatus shown in Fig. 12.4a. Put marble chips into the flask. When you are ready to collect the gas, add dilute hydrochloric acid.

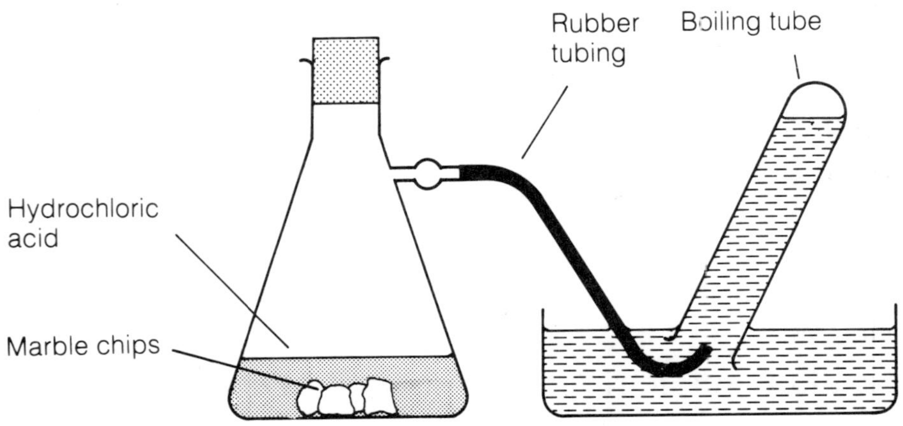

Fig. 12.4(a) Preparing and collecting carbon dioxide

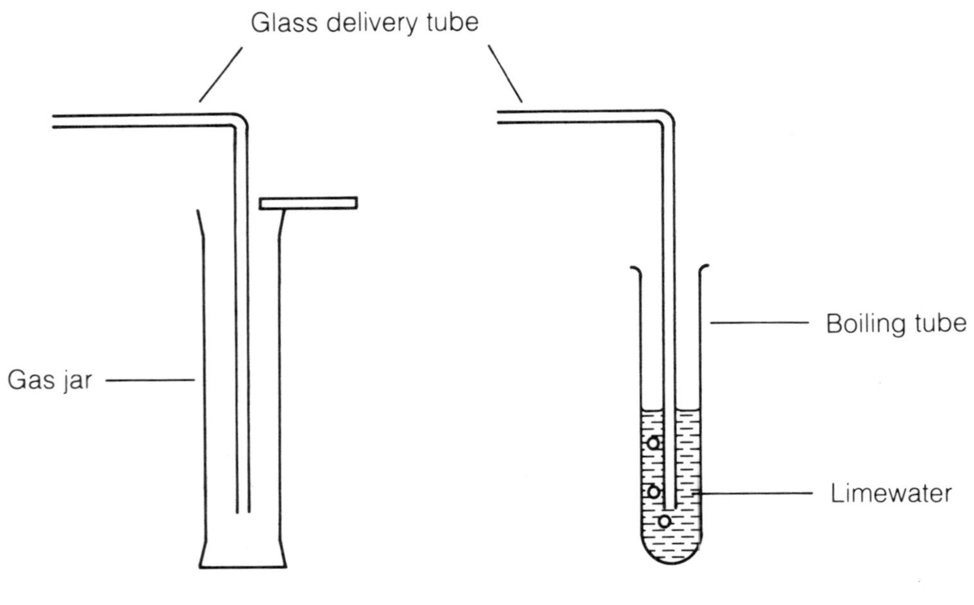

Fig. 12.4(b) Another method of collecting carbon dioxide

Fig. 12.4(c) Bubbling carbon dioxide through limewater

2 Discard the first boiling tube of gas. Collect two boiling tubes of gas over water as shown in Fig. 12.4a. Cork each tube before you take it out of the water. Stand them in a rack.

3 Connect a right-angled delivery tube as in Fig. 12.4b. Fill two gas jars with gas by downward delivery. Cover the gas jars with lids.

4 With the right-angled delivery tube attached, bubble the gas through limewater in a boiling tube (Fig. 12.4c).

5 Bubble the gas through water containing universal indicator.

6 Take one of the boiling tubes of gas collected in step **2**. Lower a burning taper into the gas.

7 Take the second boiling tube of gas, standing in a rack. Ask your teacher to demonstrate what happens when a strip of burning magnesium is lowered into the gas. **Do not hold the boiling tube in your hand; leave it in the rack**. Note the appearance of the contents of the boiling tube. Add dilute hydrochloric acid. Again, observe carefully. Filter. Can you identify the two products of the reaction between magnesium and carbon dioxide?

8 Is carbon dioxide denser or less dense than air? Figure 12.4d shows how you can find out. Put a gas jar of carbon dioxide on top of one of air. Remove the lids. Count 20 seconds. Then replace the lids. Test both jars with a lighted splint. Where is the carbon dioxide now?

9 Repeat step **8** with a gas jar of carbon dioxide underneath a jar of air (Fig. 12.4e). What have steps **8** and **9** told you about the density of carbon dioxide?

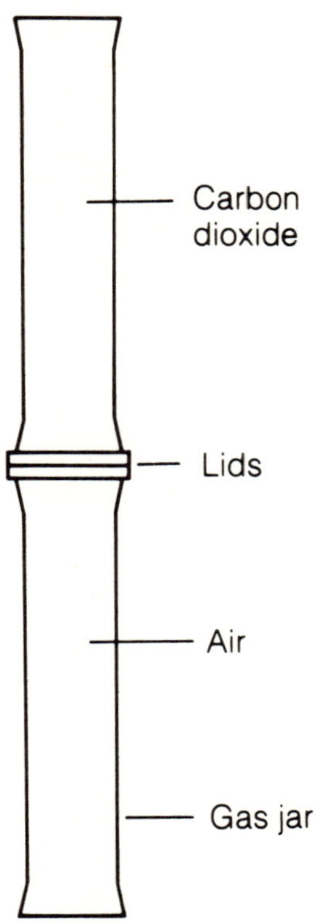

Fig. 12.4(d) Testing the density

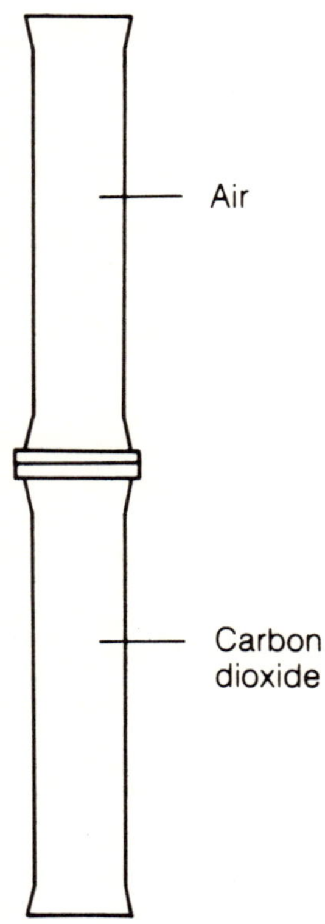

Fig. 12.4(e) Testing the density

10 Summarise the results of your observations on carbon dioxide, either in your own words or by copying and completing the following passage.

In appearance, carbon dioxide is ________ and ________.
It changed universal indicator from ________ to ________, proving that it is ________ ________. When carbon dioxide is bubbled through limewater, we saw ______________. The reason for this is ______________________. When a burning splint was put into magnesium, we saw ______________. When burning magnesium was plunged into carbon dioxide, we saw ________. Two solids were formed; in appearance they were ________ and ________.

The reaction that happens is

magnesium + carbon dioxide → ———— + ————

The chemical equation for this reaction is

Carbon dioxide is denser/less dense than air. We proved this by the experiment in which carbon dioxide rose/sank in contact with air.

Copy and complete the following passage.

The reaction which produces carbon dioxide in this preparation is

calcium carbonate + hydrochloric acid → ———— + ———— + ————

The chemical equation is

Experiment 12.5
Fire extinguishers

1 Place a metal tray on a heat-proof bench mat. Sprinkle some wood shavings on the metal tray (Fig. 12.5a). Set fire to them.

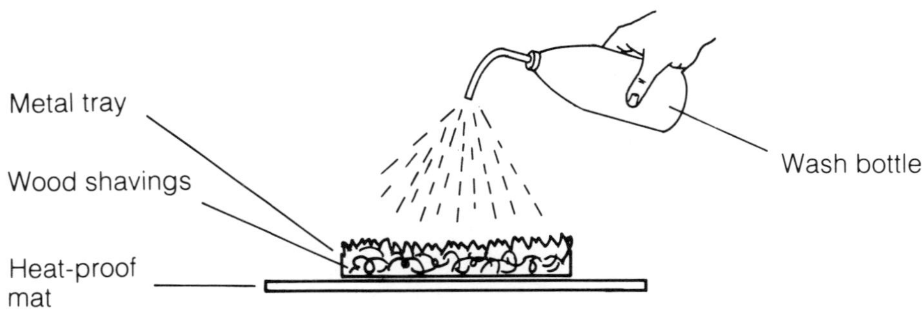

Fig. 12.5(a) Using water on a fire

2 Direct a jet of water from a wash-bottle on to the fire.

3 Light a similar fire. Sprinkle sodium hydrogencarbonate on to the fire (Fig. 12.5b).

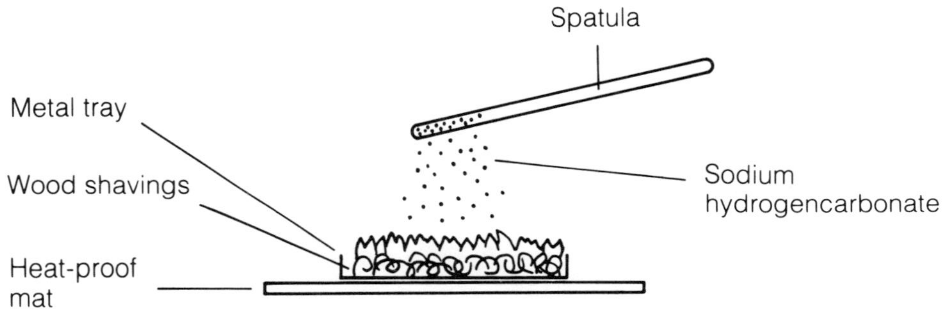

Fig. 12.5(b) A powder extinguisher

4 Assemble the model of a foam extinguisher shown in Fig. 12.5c. The conical flask contains a saturated solution of sodium hydrogencarbonate and 5 cm^3 of washing-up liquid. Be careful to lower the small test-tube of concentrated hydrochloric acid into the flask without letting it meet the solution. **Wear safety glasses.**

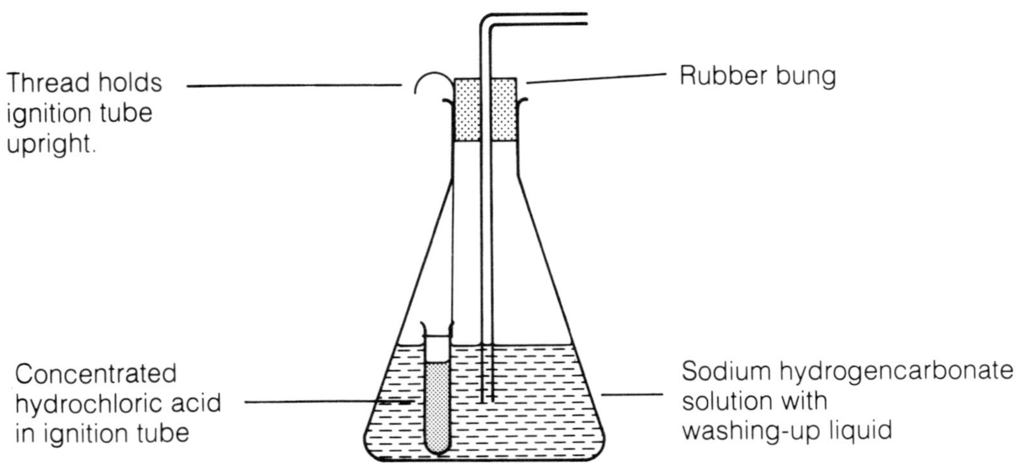

Fig. 12.5(c) A foam extinguisher

5 Again, set fire to some wood shavings in a metal tray. Shake the model fire extinguisher to let the acid meet the sodium hydrogencarbonate solution. Point the glass tube towards the fire.

6 Explain how the fire is extinguished in these three methods. Refer to the fire triangle in your explanation.

Experiment 12.6
How easily do fabrics burn?

1 Collect as many different fabrics as you can: silk, rayon, cotton, wool, nylon, polyester and others. Cut the fabrics into 10 cm squares.

2 Stand a Bunsen burner on a heat-proof mat. Light it, and then turn the flame down very low. Hold a piece of fabric in tongs over the flame. Note the way in which the fabric burns.

3 Repeat with the other fabrics in turn. List the fabrics you have tested in order of the ease with which they burn.

4 Do flame retardants work? You can make a flame retardant by dissolving 87 g of borax and 37 g of boric acid in 1 litre of water. Soak a piece of fabric in the solution. Let it dry, and then test it as in step **2**.

5 Test a commercial flame retardant, such as Proban®.

6 Does flame retardant withstand washing?
Treat a piece of fabric with flame retardant, let it dry, and then wash it in a soap powder, e.g. Lux® or Dreft®. Let it dry, and test it as in step **2**. Test the effect of washing in a soapless detergent, e.g. Tide® or Daz®.

7 Do flame retardants work equally well on all fabrics?

Experiment 12.7
Investigating shells

You need a sea shell. The experiment will enable you to say what it is made of.

1 Put a sea shell in a mortar. Using a pestle, grind it to a powder.

2 Half fill a test-tube with water. Add a little powdered shell. Does it dissolve?

3 Pour 2 cm depth of hydrochloric acid into a test-tube. Add one spatula measure of powdered shell. Does the shell dissolve or react?

4 If a gas is formed in **3**, collect it (see Fig. 12.7a). Collect three corked test-tubes of gas.

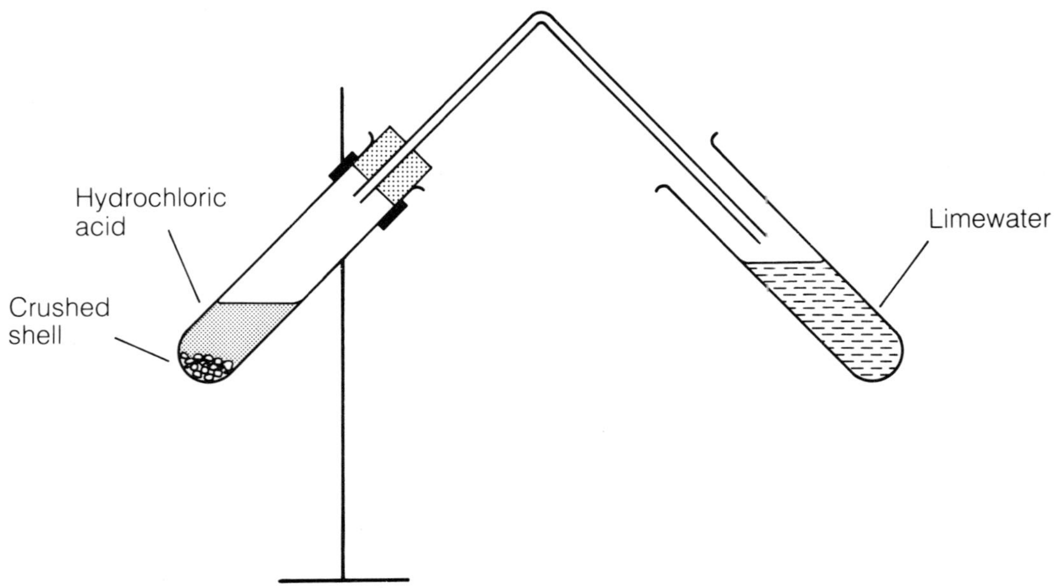

Fig. 12.7(a) Acid and shell

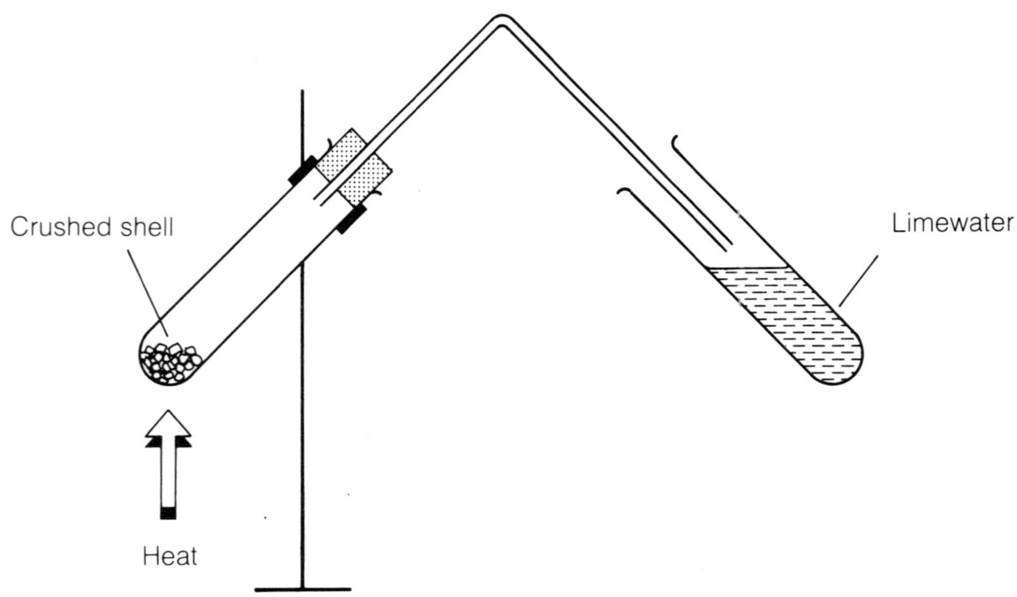

Fig. 12.7(b) Heating powdered shell

5 Into one test-tube, put a glowing splint.
Into another test-tube, put a lighted splint.
To the third, add 2 cm^3 of limewater, and shake.
From your results, say what the gas might be.

6 Find out what happens when some shell is heated. Put two spatula measures of powdered shell into the apparatus shown in Fig. 12.7b. Heat. Collect four test-tubes of gas.

7 Test the four test-tubes of gas with
- **a** universal indicator solution (three drops),
- **b** a glowing splint,
- **c** a lighted splint.
- **d** limewater (2 cm^3).

Say what you think the gas might be.

8 Let the residue from **7** cool. Then tip it out of the test-tube on to a watch glass. Touch it with damp universal indicator paper. What have you found out about the nature of the residue? What kind of compound is it? The results of step **7** and step **8** should give you a good idea of the kind of compound shells are made of.

9 Now do a test to see what metal is present in the shell. Hold a piece of shell in tongs. Dip it into hydrochloric acid. Hold it in a Bunsen flame. Note the colour.

10 Some metals and their compounds will colour a flame. Here are some examples:
sodium: yellow,
potassium: lilac,
copper: blue–green,
calcium: brick red,
barium: apple green,
strontium: scarlet.
Which of these metals have you detected? Is it present as a metal or as a compound?

11 Put together the information you have found out as a result of your experiments. What is the chemical in sea shells?

TOPIC 13 Sulphur and its compounds

Experiment 13.1 Crystals of sulphur

Method 1

1 Put two spatula measures of powdered sulphur into a boiling tube.

2 Add 20 cm^3 of 1,1,1-trichloroethane. An alternative solvent is methylbenzene (toluene). (These liquids are covalent solvents. **They are flammable; take care.**)

3 Stand a beaker of water on a tripod and gauze. Heat the water to boiling point.

4 Switch off the Bunsen burner. Put the boiling tube to stand in the beaker of hot water (see Fig. 13.1a). Shake the mixture from time to time over a period of 10 minutes.

5 Filter the solution into a warm boiling tube (Fig. 13.1b). If some crystals form, put the boiling tube into the beaker of hot water. This will make the crystals dissolve again.

6 Take the boiling tube out of the hot water. Leave the solution to stand.

7 When crystals form, filter. Examine the crystals through a hand lens. Make a drawing to show the shape of the crystals.

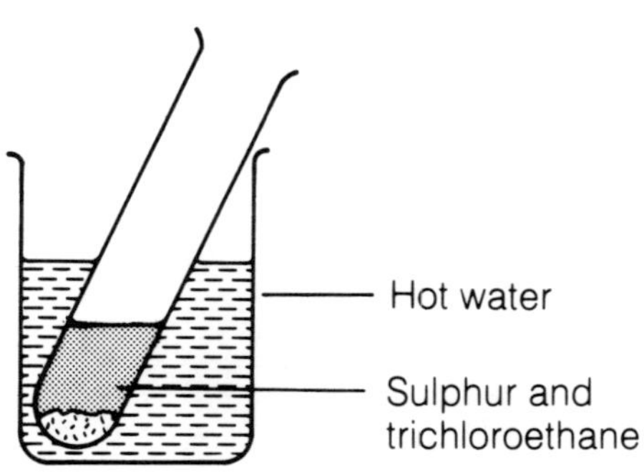

Fig. 13.1(a) Warming sulphur with a solvent

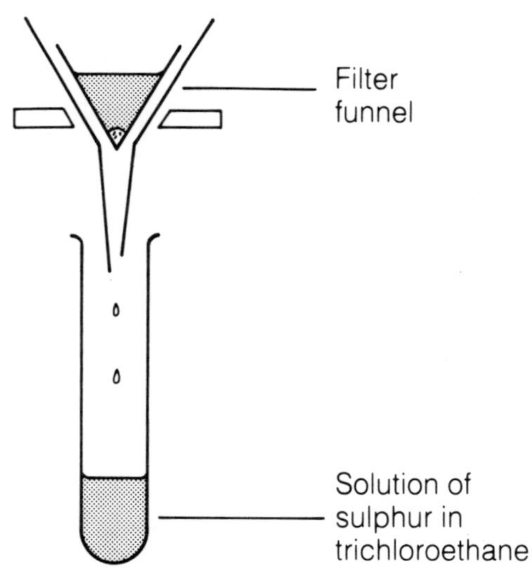

Fig. 13.1(b) Filtering the solution

Method 2

1 Pour a 3 cm depth of dichloromethane into a boiling tube. **This is a flammable solvent; take care**.

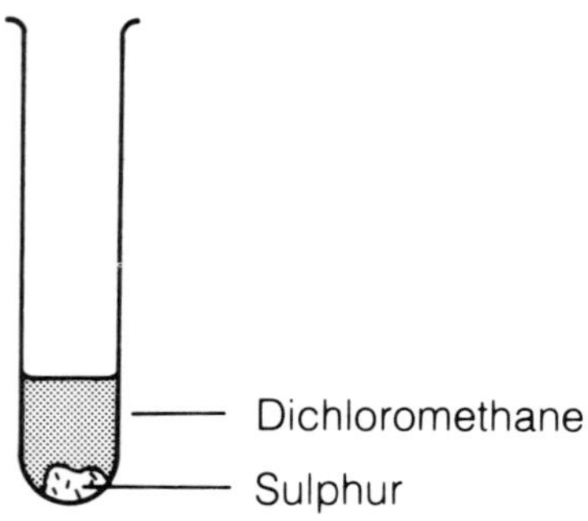

Fig. 13.1(c) Making a solution

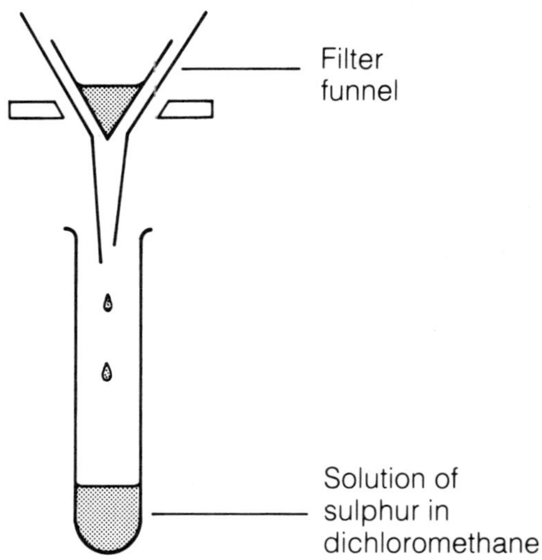

Fig. 13.1(d) Filtering

2 Add a little powdered sulphur, and shake (Fig. 13.1c). Add more sulphur. Stop when you have a saturated solution.

3 Filter the solution into a boiling tube (Fig. 13.1d).

4 Leave the solution to stand in a fume cupboard.

5 Sulphur crystals will form when the solvent evaporates. Look at them with a hand lens. Make a careful drawing. Compare them with the crystals you obtained by method 1. Can you explain the difference?
Note: You can leave the solution to crystallise in a crystallising dish. If you leave the solution to stand in a boiling tube, it will take longer to crystallise, but you will obtain larger crystals.

Experiment 13.2
Finding out what is formed when sulphur burns

1 Place some powdered sulphur on a combustion spoon. Have ready six boiling tubes in a rack and corks to fit them.

2 Carefully heat the spoon in a moderate Bunsen flame until the sulphur starts to burn. Note the colour of the flame. Do not try to smell the gas produced.

3 Lower the burning sulphur into a boiling tube, which is standing in a rack. **Do not hold the boiling tube in your hand**. After 5 seconds, remove the combustion spoon, and cork the tube. Collect six tubes of gas in this way (Fig. 13.2).

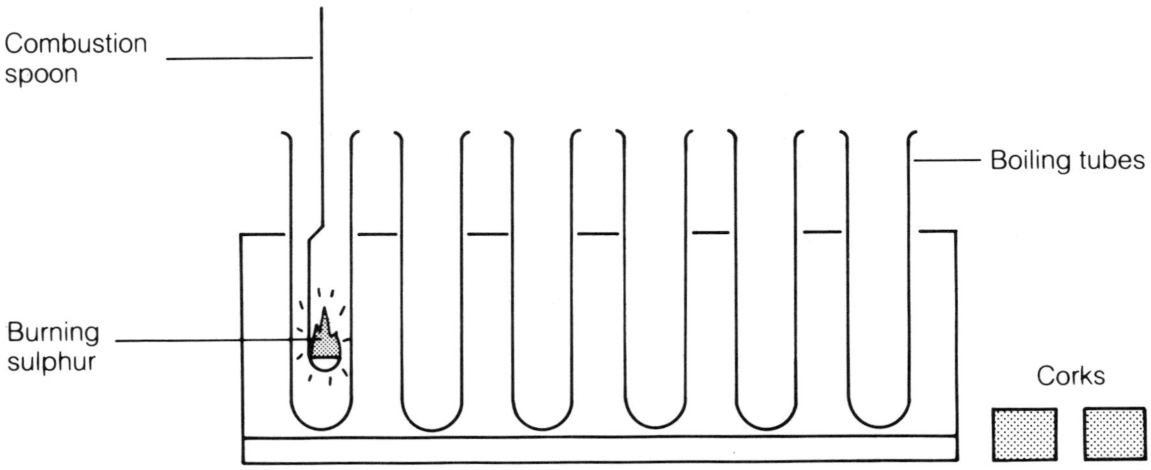

Fig. 13.2 Burning sulphur

4 The next step is to add a different substance to each boiling tube of gas. Add 1 cm^3 of each solution, replace the cork and shake. Record your observations in a table.

Table Reaction of the gas formed when sulphur burns

Substance	*Colour*	*Colour after shaking with gas*
Litmus solution		
Potassium manganate(VII) solution		
Potassium dichromate(VI) solution		
Piece of apple peel		
Green leaf		
Flower petal		

Copy and complete the following passage.

Sulphur burns to form the gas sulphur dioxide, SO_2. The gas looks ———— and smells ————.
The action of the gas on litmus shows that sulphur dioxide is ————.
Potassium manganate(VII) and potassium dichromate(VI) are oxidising

agents. You know that they react with sulphur dioxide because
——————————.

When a solution of an oxidising agent gives oxygen to sulphur dioxide,

sulphur dioxide	+ water	+ oxygen from oxidising agent	→	————
$SO_2(aq)$	+ $H_2O(l)$	+ [O]	→	————

The product of oxidation is ————.
Sulphur dioxide can ———— some plant pigments.

Experiment 13.3
Alum: a useful sulphate

Part A

Alum is aluminium potassium sulphate, $KAl(SO_4)_2{\cdot}12H_2O$. (See Experiment 1.2.)

1 Take a 250 cm^3 beaker. Add 100 cm^3 of water. Add alum and stir. Continue until you have a saturated solution.

2 Soak a piece of white cotton cloth in the solution for 1 minute.

3 Lift out the cotton, and wring it.

4 Pour 100 cm^3 of bench sodium carbonate solution into a 250 cm^3 beaker. Boil it.

5 Drop the piece of cotton cloth into the boiling solution. After 1 minute, lift it out with tongs.

6 Rinse the cloth.

7 Prepare 200 cm^3 of a solution of a dye such as Congo red, eosin or alizarin. Divide the solution into two parts. Into one portion, put an untreated piece of cloth. Into the other portion, put the piece of cloth which has been treated with alum. Leave both for 1 minute.

8 Lift out the pieces of cloth. Rinse them with water. Lay the pieces of cotton side by side on a white tile. Do you notice any difference in colour?

9 Alum is used in dyeing as a *mordant*. To understand how it works, you should do Part B.

Part B

1 Dissolve 1 g of alum in 50 cm^3 of water.

2 Dissolve a small quantity (0.1 g) of the dye Congo red (or eosin or alizarin) in sodium hydroxide solution. Add it to the solution in step **1**.

3 Add bench sodium carbonate solution. A precipitate forms.

4 Filter. You can see that the dye is in the precipitate. Try to wash it out with water.

5 What is the precipitate that forms when alum and sodium carbonate are mixed? Remember that sodium carbonate solution is alkaline:

aluminium salt + alkali → ————

What forms in the fibres of the cloth when alum and sodium carbonate solutions meet?
How does this precipitate react with the dye?
How does the reaction explain the use of alum in dyeing?

Part C *Alum as a flame retardant*

Many synthetic materials burn much faster than natural fabrics, such as wool and silk. Often, synthetic fabrics are soaked in a chemical which will slow down the rate of burning: a flame retardant. Alum is used as a flame retardant.

1 Take a piece of filter paper, and draw a pencil line down the middle. Soak half the paper in a solution of alum. Let the paper dry.

2 Set fire to the filter paper (see Fig. 13.3a). Describe the difference in the way the two halves burn.

3 It is really more important to find out how clothing burns. Make a collecton of fabrics: wool, cotton and as many different synthetic fabrics as you can find. Cut each piece of fabric into two.

4 Soak one half of each fabric in alum solution. Dry these pieces in an oven.

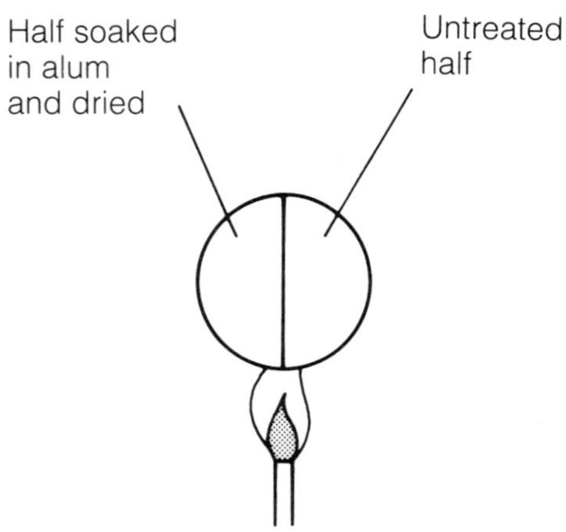

Fig. 13.3(a) Burning a filter paper

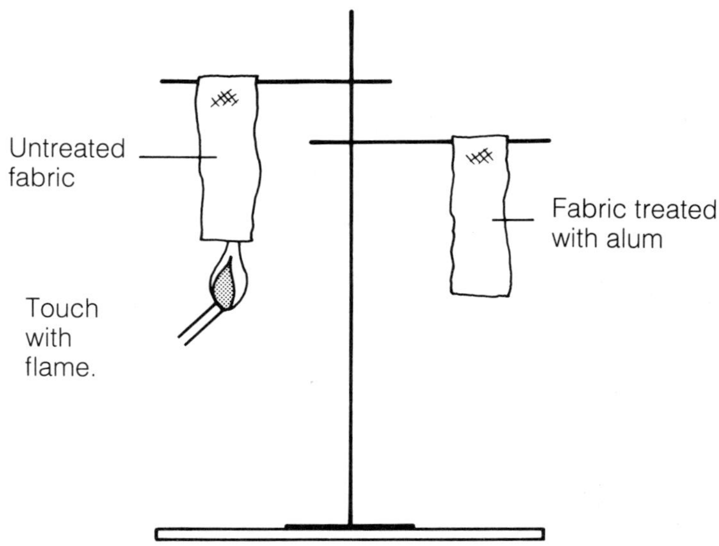

Fig. 13.3(b) Burning fabric

5 Suspend two fragments of the same fabric, the one which you have soaked in alum and the untreated fragment (see Fig. 13.3b). Have a heat-proof mat on the bench. Set fire to the two strips of material. Watch carefully to see if there is a difference in the way they burn.

Experiment 13.4
Sodium hydrogensulphite: some uses

A permanent crease for wool

(This is an experiment that you can do at home.)

1 Make a solution of 3 g of sodium hydrogensulphite, $NaHSO_3$, in 100 cm^3 of water.

2 Take two samples of woollen material. Mark one of them by tacking a piece of thread through it.

3 Absorb some sodium hydrogensulphite solution in a sponge. Sponge a line of solution down the centre of the notched wool sample. Crease the sample along the sponged line. Press in the crease with a steam iron for 30 seconds.

4 Use the steam iron to press a crease in the untreated sample.

5 Immerse both samples in warm water (about 70 °C) containing a little detergent.

6 Rinse the samples, wring them, and hang them up to dry. Has the treatment produced a permanent crease?

Preserving fruit

(This is an experiment that you can do at home.)

Sulphur dioxide is used as a food preservative. The additives E220 to E227 are compounds which release sulphur dioxide slowly. Sodium hydrogensulphite is one of them. Sodium metabisulphite, which is widely used in home-brewing, is another.

1 *Does sulphur dioxide preserve fruit?* Your problem is to design an experiment to find out.

2 You are supplied with sodium hydrogensulphite or sodium metabisulphite or home-brew Campden tablets.

3 You will have to find two pieces of fruit, e.g. two apples. Sulphur dioxide is formed slowly when sodium hydrogensulphite reacts with water. You will have to find a way of keeping the sulphur dioxide in contact with the fruit.

4 Sketch the apparatus you will use. Explain what control experiment you need to do.

5 If you can find a suitable set-up, go ahead with your experiment.

Experiment 13.5
Invisible writing

Dilute sulphuric acid can be used as invisible ink.

1 Pour a little dilute sulphuric acid into a small beaker. Dip a brush into the dilute sulphuric acid, and write a message on a piece of filter paper. (If you can't find a brush, use a glass rod.)

2 Let the writing dry. You now have an invisible message.

3 Hold the filter paper carefully some distance above a Bunsen flame. Your message will become visible.

4 Paper contains cellulose, which is a carbohydrate, that is, a compound of carbon with hydrogen and oxygen in the ratio of two hydrogen atoms to one oxygen atom. It has the formula $C_m(H_2O)_n$.

- **a** What are the letters you can see made of?
- **b** What reaction has occurred between sulphuric acid and cellulose?
- **c** Why did you have to heat to make this reaction happen?

TOPIC 14 Nitrogen compounds

Experiment 14.1
Heating protein foods with soda–lime

1 Mix half a spatula measure of cheese with one spatula measure of soda–lime. (Soda–lime is a mixture of a strong base and a drying agent.) Put the mixture into a test-tube.

2 Take a drying tube like that shown in Fig. 14.1. Push a small wad of glass wool into the bottom of the tube, and then fill it with chips of activated charcoal. This is an *adsorbent* which will trap most of the smells produced in the experiment.

3 Set up the apparatus as in Fig. 14.1.

4 Heat the test-tube. At the open end of the tube, smell any gas which is given off. Test it with both damp blue litmus paper and damp red litmus paper.

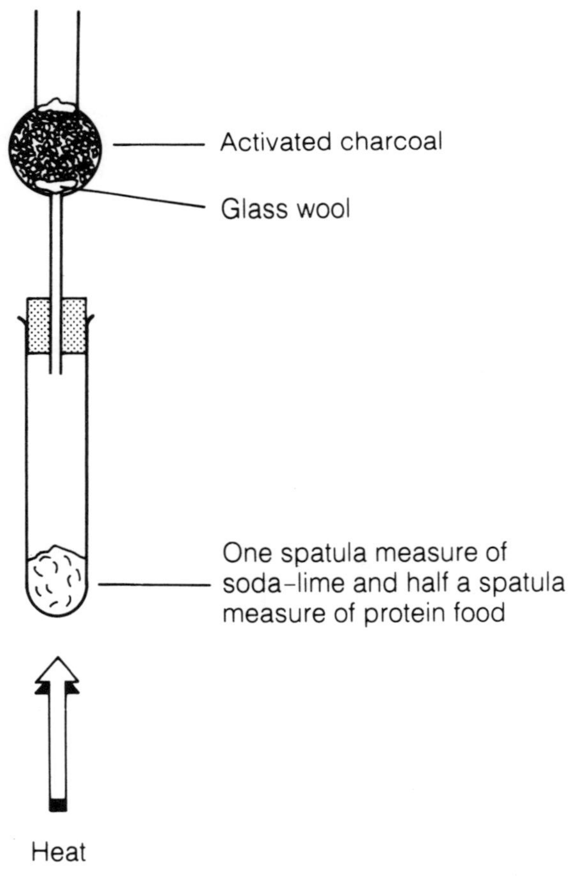

Fig. 14.1 Heating a protein food with soda–lime

5 Repeat the experiment with fish or meat.

6 What is the gas that is formed when protein foods react with soda–lime?

The formation of this gas proves that proteins must contain a certain element. Which element is this?

7 The information you have gained is used to help plants to build proteins. They are fertilised with compounds containing this essential element. What sort of compounds contain this element? How are these compounds made in industry?

Experiment 14.2
Making the fertiliser ammonium sulphate

Ammonium sulphate is used as a fertiliser. It is made industrially from ammonia produced by the Haber process and sulphuric acid produced by the contact process. The reaction is an example of *neutralisation*:

ammonia + sulphuric acid → ammonium sulphate

$2NH_3(aq) + H_2SO_4(aq) \rightarrow (NH_4)_2SO_4(aq)$

1 Take 50 cm of dilute sulphuric acid in a beaker (Fig. 14.2a).

2 With a teat pipette, add ammonia solution. Stir with a glass rod. Take out a drop of the solution, and dab it on to a piece of purple litmus paper (Fig. 14.2b). The paper will turn red, showing that the solution is still acidic.

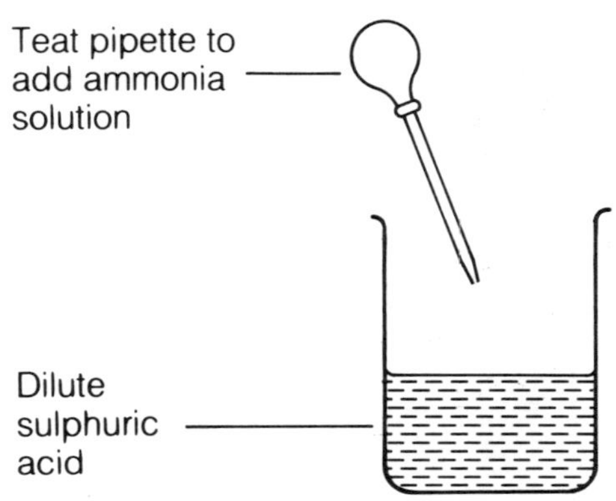

Fig. 14.2(a) Adding ammonia

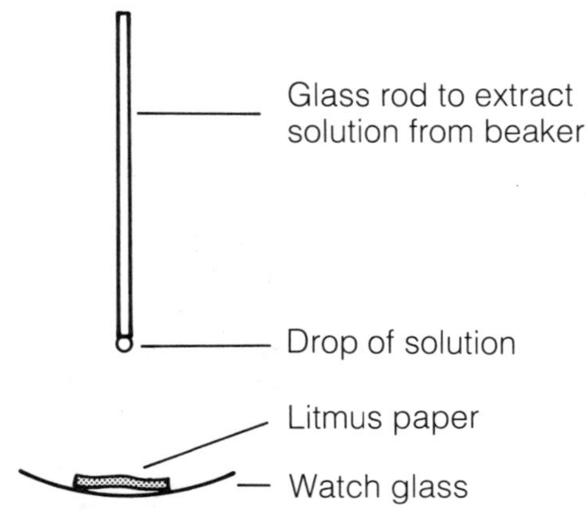

Fig. 14.2(b) Testing with litmus

3 Add more dilute ammonia solution, stir, take out a drop of solution and test it as before. Continue adding more ammonia until the litmus turns blue, showing that the solution is alkaline. All the acid has been neutralised, and there is an excess of ammonia present. This does not matter as the excess will be driven off as ammonia gas and steam when you begin to heat the solution.

4 Pour the solution into an evaporating basin. Put it on to a water bath. From time to time, dip a cold glass rod into the solution. When crystals form on the rod, you know that the solution is sufficiently concentrated to crystallise. Take the evaporating basin off the water bath, and leave it to stand. Collect the crystals of ammonium sulphate formed.

5 Save your product to test in Experiment 14.5.

Experiment 14.3
Making ammonia

1 Mix a spatula measure of ammonium sulphate with a spatula measure of calcium hydroxide.

2 Put the mixture into a test-tube. Warm it gently.

3 Do the following tests on the gas which is evolved.

a Smell it cautiously (Fig. 14.3a).

b Hold a piece of damp red litmus paper and a piece of damp blue litmus paper in the stream of gas (Fig. 14.3b).

c Unstopper a bottle of concentrated hydrochloric acid. Hold your test-tube over the open bottle for a few seconds (Fig. 14.3c).

d Put a spot of copper(II) sulphate solution on to a piece of filter paper. Hold the filter paper in the stream of gas.

Fig. 14.3(a) How to smell a gas

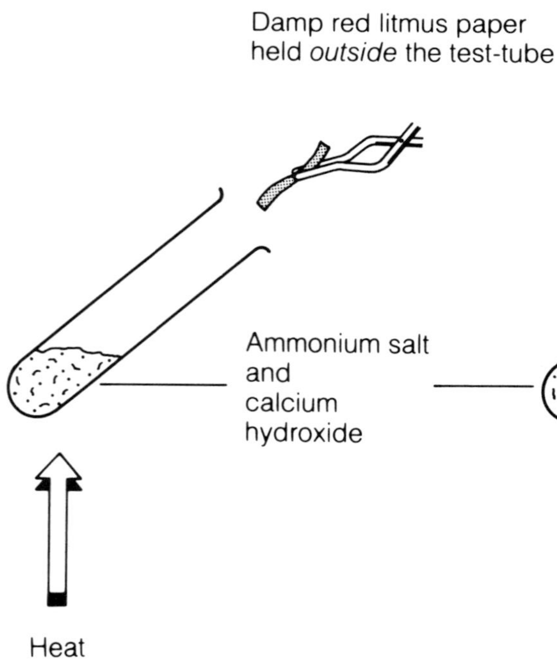

Fig. 14.3(b) How to test with litmus

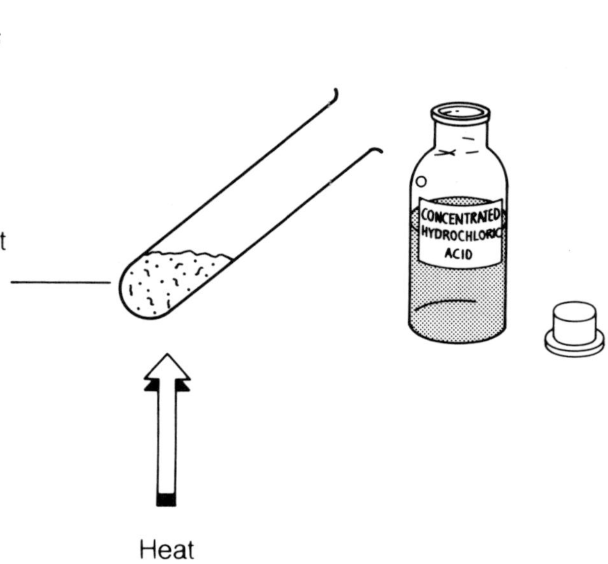

Fig. 14.3(c) Testing with hydrogen chloride

4 Summarise your observations. Either write an account in your own words or copy and complete this passage.

Ammonia is formed when an ammonium salt is heated with a strong base. The salt we used was ————, and the strong base was ————. The word equation for the reaction is

$$\begin{matrix}\text{ammonium}\\ \text{sulphate}\end{matrix} + \begin{matrix}\text{calcium}\\ \text{hydroxide}\end{matrix} \rightarrow \text{ammonia} + \begin{matrix}\text{calcium}\\ \text{sulphate}\end{matrix} + \text{water}$$

The chemical equation is

————————————————————————————

Ammonia is a gas with a ———— smell. It turns ———— litmus ————, showing that it is ————. The gas hydrogen chloride escapes from an open bottle of concentrated hydrochloric acid. When it meets ammonia, we see ————. This substance is ————.

When ammonia reacts with copper salts we see ————. This reaction can be used as a test for ammonia.

5 Farmers use ammonium salts as fertilisers. They treat very acidic soils with 'lime'.

a Why are ammonium salts used as fertilisers?

b What is the chemical name for 'lime'? Why is it used on very acidic soils?

c Is there any disadvantage in applying both ammonium salts and lime at the same time?

Experiment 14.4
Making the fertiliser ammonium phosphate

Plants need nitrogen, phosphorus and potassium. The fertiliser ammonium phosphate contains two of these elements, and, unlike most phosphates, it is soluble. It is made in huge quantities by ICI. They import the ore rock phosphate, which contains calcium phosphate(V). They make ammonia by the Haber process. From these two starting materials, they make ammonium phosphate. Sulphuric acid is vital to the process. Sulphuric acid drives phosphoric acid out of calcium phosphate. Later in the process, phosphoric acid is neutralised by ammonia.

1 Weigh out 10.1 g of powdered rock phosphate. If you do not have the ore, weigh out 9.3 g of calcium phosphate(V).

2 Put the phosphate into a 100 cm^3 beaker. Add 30 cm^3 of water.

3 Add 40 cm^3 of bench (2 M) sulphuric acid. Stir for 10 minutes.

4 Filter. The residue is calcium sulphate. The filtrate is a solution of phosphoric acid.

calcium phosphate + sulphuric acid → calcium sulphate + phosphoric acid

5 To the filtrate from step **4**, add 5 cm^3 of bench ammonia solution. Stir. Take a drop of the solution on a glass rod, and place it on a strip of purple litmus paper. It will turn red.

6 Repeat step **5** until the litmus turns blue, showing that all the phosphoric acid has been neutralised by ammonia.

phosphoric acid + ammonia → ammonium phosphate

7 Transfer the solution to an evaporating basin. Evaporate on a water bath until crystals form. You will need to evaporate almost to dryness. These crystals are ammonium phosphate.

8 The instructions in steps **1** and **3** are precise. What would happen if you added an excess of sulphuric acid? How would it affect your product? Would it matter if you used less sulphuric acid than is needed to react with the phosphate you weighed out?

9 Save your product to test in Experiment 14.5.

Experiment 14.5
Testing the fertilisers that you have made

In this experiment, you are asked to find out whether ammonium sulphate and ammonium phosphate affect the growth of plants from seed.

1 Decide what kind of seed you are going to plant, and obtain some.

2 Decide what kind of containers you are going to plant seed in, and obtain some.

3 Decide whether you are going to grow seed in sand or soil or cotton wool or some other substance. Obtain some of the substances you choose.
If you choose sand or soil, you should wash it first. What is the reason for this? What do you need to wash out?

4 Make up 1 g/l solutions of ammonium sulphate and ammonium phosphate. You will have to water one batch of seed with ammonium sulphate solution and another with ammonium phosphate solution. What should you do as a control experiment?

5 Think about how you can make the test fair. How much seed should you use? How much sand or soil should you use? How much ammonium salt solution should you use?

6 Think about how you can measure growth.

7 Write a plan for your investigation.

- Say what seed you plan to use, what containers you can obtain and what growing medium you decided on.
- Say what you are going to do to make sure that the test is fair. In scientific experiments, you can alter only one thing at a time; otherwise, you will not be able to make sense of the results.
- Explain how you will measure growth.
- Include experiments which you may do with other fertilisers, e.g. ammonium nitrate.
- When you have a written plan, show it to your teacher, and obtain permission to start your investigation.

Experiment 14.6
Two ammonium chloride problems

Problem 1

1 Take a strip of copper. Hold it in tongs over a Bunsen flame. The outside of the copper will be converted into copper(II) oxide, which is a black solid.

2 Put half a spatula measure of ammonium chloride on to the strip of blackened copper. Warm it *gently*.

3 Make a note of what you see. You should observe *three* things.

4 Explain the chemical reactions that have occurred.

ammonium chloride $\xrightarrow{\text{heat}}$ ————————

———————— + copper(II) oxide $\xrightarrow{\text{heat}}$ ————————

Problem 2

1 Set up the apparatus shown in Fig. 14.6. The strip of damp universal indicator paper lies inside the mouth of the test-tube.

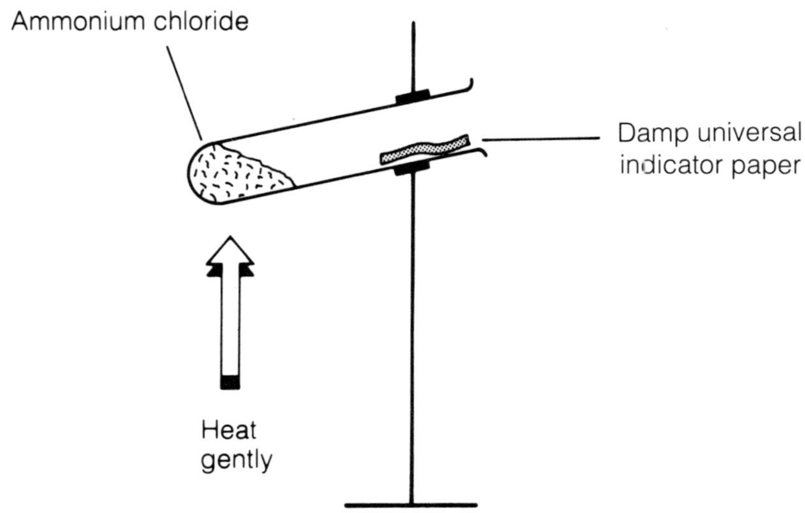

Fig. 14.6 Heating ammonium chloride

2 Heat the ammonium chloride *gently*.

3 Explain your observations.

- What happens to the universal indicator paper?
- What causes the change or changes?
- What else happens at the open end of the test-tube?
- Write a chemical equation for the change.

TOPIC 15 Halogens

Experiment 15.1
Investigating some reactions between halogens

1 Take 5 cm^3 of chlorine water in a test-tube. Add 1 cm^3 of trichloroethane, and shake. Note the appearance of both the water layer and the trichloroethane layer. (Another organic solvent which can be used instead of trichloroethane is hexane.)

2 Repeat with bromine water.

3 Iodine does not dissolve in water. You can dissolve iodine in potassium iodide solution. Repeat step **1** with this solution.

4 Make a copy of Table A. Enter the colours of the solutions of chlorine, bromine and iodine in water and in trichloroethane (or hexane).

Table A Colours of solutions of the halogens

Halogen	*Colour of solution in water*	*Colour of solution in trichloroethane*
Chlorine		
Bromine		
Iodine		

5 Take 5 cm^3 of potassium bromide solution. Add 1 cm^3 of chlorine water, and shake. Add 1 cm^3 of trichloroethane, and shake. Make a note of what you see (Table B). Deduce what chemical reaction has happened.

6 Take 5 cm^3 of potassium iodide solution. Add 1 cm^3 of chlorine water, and shake. Add 1 cm^3 of trichloroethane, and shake. Make a note of what you see (Table B). Deduce what chemical reaction has happened.

Table B Results

Halogen solution	*Salt solution*	*Observation*	*Deduction*
Chlorine	Potassium bromide		
Chlorine	Potassium iodide		

7 Try to predict what will happen if you add bromine water to potassium iodide solution:

bromine + potassium iodide solution → ————

8 Test your prediction.

9 Copy and complete these equations:

chlorine + potassium bromide → __________

chlorine + potassium iodide → __________

bromine + potassium iodide → __________

10 Your results give you information about the reactivity of the halogens.
Which is the more reactive (the more keen to form compounds):
chlorine or bromine?
Which is the more reactive: chlorine or iodine?
Which is the more reactive: bromine or iodine?
List the halogens in order of reactivity.

Compare your list with the Periodic Table. What is the resemblance or difference between the order you have found and the order in the Periodic Table? You will see two other halogens listed in the Periodic Table. Predict where they would come in the order of reactivity which you have found. Explain why these two halogens are not used in school experiments.

Experiment 15.2
Iodine gets another chance

Iodine and sulphur are reactive non-metallic elements. Which do you think would win in a competition between iodine and sulphur?

1 First, the reaction between zinc and sulphur to form zinc sulphide must be **demonstrated by the teacher** in a fume cupboard (see Appendix).

2 Take a spatula measure of zinc sulphide in a boiling tube. Add 10 cm^3 of water, and shake. The solid will not dissolve; it will form a suspension.

3 Add 10 cm^3 of a solution of iodine in potassium iodide solution (Fig. 15.2a).

4 Warm the mixture gently over a Bunsen flame (Fig. 15.2b). Leave it to stand for 5 minutes, shaking from time to time.

5 You will see two changes. Note what happens to the solution and to the solid. What is the solid now? What reaction has taken place in the solution? Copy and complete the equation

zinc sulphide + iodine → ———— + ————

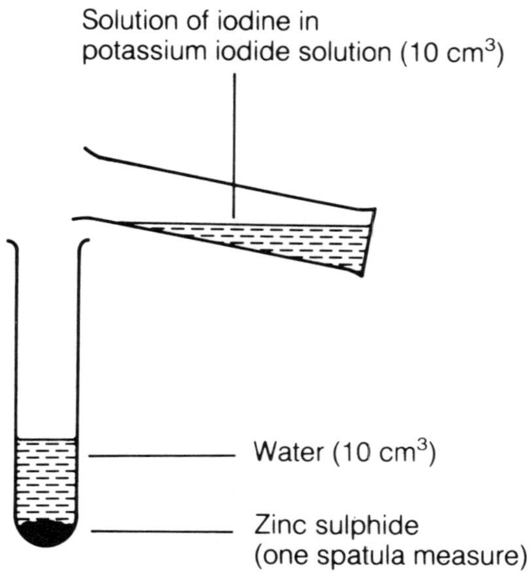

Fig. 15.2(a) Adding iodine to zinc sulphide

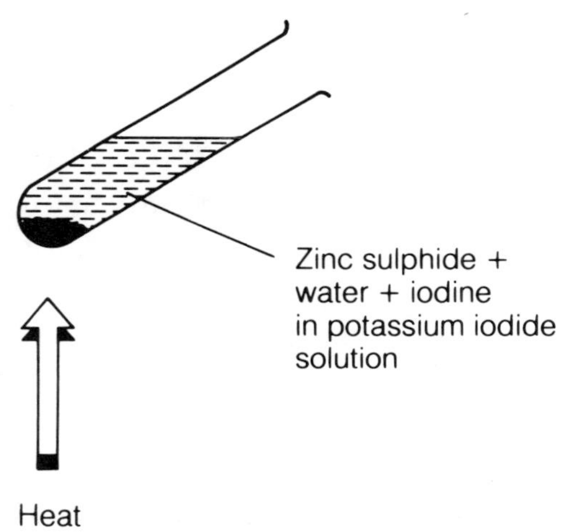

Fig. 15.2(b) Warming the mixture

Experiment 15.3
Comparing household bleaches

Household bleaches contain the compound sodium chlorate(I), NaClO. It reacts with acids to give chlorine. Chlorine is the substance which does the actual bleaching. The amount of chlorine which a bleach supplies can be measured. The chlorine produced is allowed to react with potassium iodide to give iodine. The amount of iodine produced can be found by titration. A solution of sodium thiosulphate(VI), $Na_2S_2O_3$, reacts with iodine.

1 First, make up solutions of three or more household bleaches. **It is better if the teacher does this part of the work**. Take 100 cm^3 of bleach in a measuring cylinder. Make it up to 1 litre with water. Repeat with the other two bleaches.

2 With a measuring cylinder, take 25 cm^3 of bleach A. Pour it into a conical flask.

3 With a second measuring cylinder, add 10 cm^3 of potassium iodide solution (a 10% solution).

4 With a third measuring cylinder, add 10 cm^3 of bench sulphuric acid. (If you can find only one measuring cylinder, you will have to wash it carefully between solutions.)

5 Fill a burette with sodium thiosulphate solution (a 5% solution containing a little sodium hydroxide to stabilise it). Read the burette.

6 Slowly add sodium thiosulphate solution to the conical flask. Swirl the flask as you titrate. The colour of iodine will change gradually from brown to pale yellow. When it is pale yellow, add 1 cm^3 of starch solution. Starch and iodine react to form a dark blue compound. Continue the titration until the blue colour disappears. Read the burette.

7 Write down the volume of sodium thiosulphate solution added.

8 Repeat the experiment with bleach B and then with bleach C.

9 The stronger the bleach is, the more iodine it will displace, and the larger will be the volume of thiosulphate. By comparing the titration volumes with the prices, you can find out which bleach is the best value for money.

Table Comparison of household bleaches

Bleach	*Volume of bleach in bottle in cm^3*	*Price in p*	*Price of 100 cm^3*	*Volume of sodium thiosulphate solution in cm^3*
A				
B				
C				

10 Use the figures in the last two columns to tell you which bleach is the best buy.

TOPIC 16 Hydrocarbons

Experiment 16.1
Obtaining fractions from crude oil

The petroleum industry uses crude petroleum oil as a source of petrol, diesel oil, lubricating oil and other important products. The method used to separate crude oil into these parts is called *fractional distillation*. In this experiment, you can do a fractional distillation of crude oil.

1 Set up the apparatus shown in Fig. 16.1. Put a little rocksil in the bottom of the side-arm tube. Add 2 cm^3 of crude oil.

2 Make sure that the delivery tube fits right up to the side-arm of the test-tube so that the rubber connecting tube is not exposed to vapour.

3 Make sure that the thermometer bulb is at the height of the side-arm tube, as shown in Fig. 16.1.

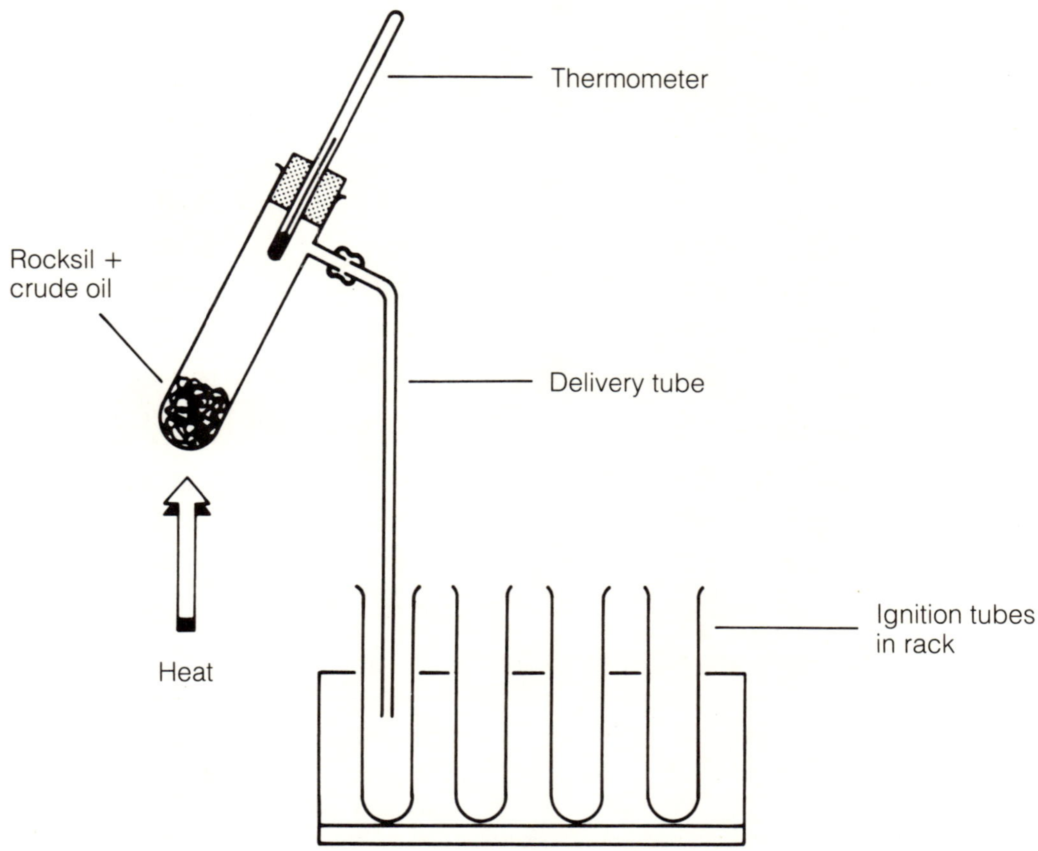

Fig. 16.1 Fractional distillation of crude oil

4 Heat the side-arm tube carefully with a blue flame 2 cm high. Collect the fraction that boils between room temperature and 70 °C. Then change the receiver.

5 Use the four ignition tubes to collect four fractions with these boiling point ranges:
tube 1, 20–70 °C,
tube 2, 70–120 °C,
tube 3, 120–170 °C,
tube 4, 170–220 °C.

6 With each of the four fractions, try the following tests.
a Pour the contents of each tube on to a watch glass. Notice the *viscosity* of each fraction, that is, the ease or difficulty with which it flows. A liquid which flows with difficulty is described as *viscous*.
b Try to light each liquid on its watch glass. Does it burn quickly or slowly? Does it have a luminous flame (giving out light) or a smoky flame?

7 Rank the fractions in order of
a increasing viscosity,
b increasing flammability.

8 Do you think each fraction is a pure substance? Give a reason for your answer.

9 Refer to a textbook to find out how many fractions the oil industry obtains from crude oil.

Experiment 16.2
Experiments with alkanes

Hexane, $CH_3CH_2CH_2CH_2CH_2CH_3$, is an alkane which is liquid at room temperature. Candle wax is a mixture of solid alkanes.

1 Does hexane dissolve in water? Shake 3 cm^3 of hexane with 3 cm^3 of water in a test-tube.

2 Is it denser or less dense than water?

3 Find out whether the following substances will dissolve in hexane:
- **a** sodium chloride,
- **b** sodium hydrogencarbonate,
- **c** margarine,
- **d** citric acid.

Add a small quantity of each separately to 3 cm^3 of hexane in a test-tube. Shake, and note the results.

4 Will hexane burn? Pour 3 cm^3 of hexane on to a watch glass. Set it alight. Note the appearance of the flame and the speed with which it burns. Compare it with crude oil.

5 What combustion products are formed when a candle burns? Set up the apparatus shown in Fig. 16.2a. The suction pump draws the combustion products through the apparatus. What changes do you see? What are the combustion products?

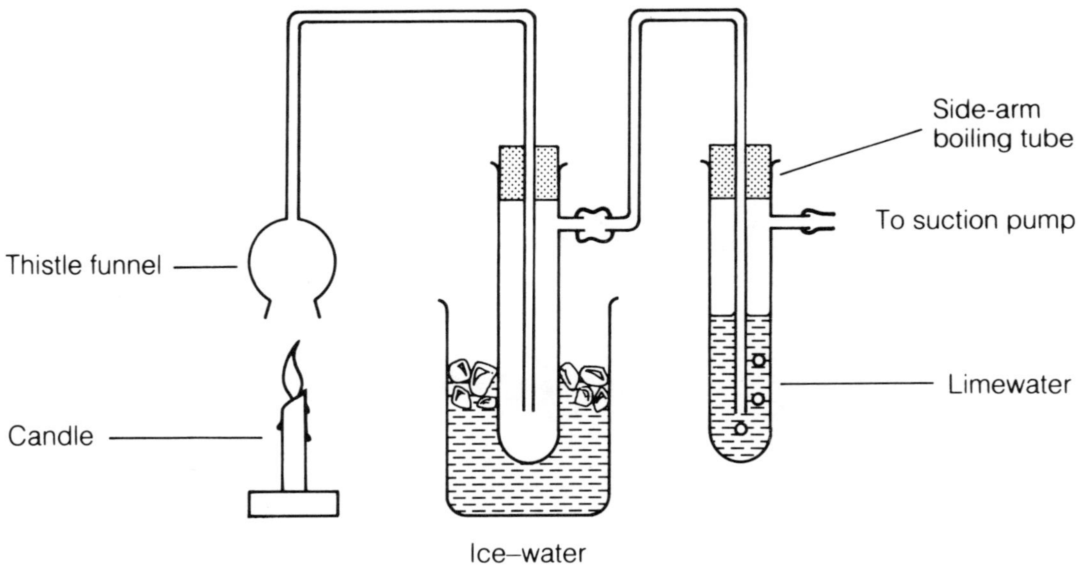

Fig. 16.2(a) Combustion products of candle wax

6 Are the oxidation products of candle wax the same as the products of combustion? You can warm 1 g of candle wax with 3 g of copper(II) oxide in the apparatus shown in Fig. 16.2b. Warm gently: you do not want the wax to vaporise before it has had a chance to react with copper(II) oxide. What change do you see in the limewater? Explain the cause of the change.

What condensate do you see in the test-tube? Test the condensate with cobalt chloride paper. Explain the result.

What are the oxidation products of candle wax?

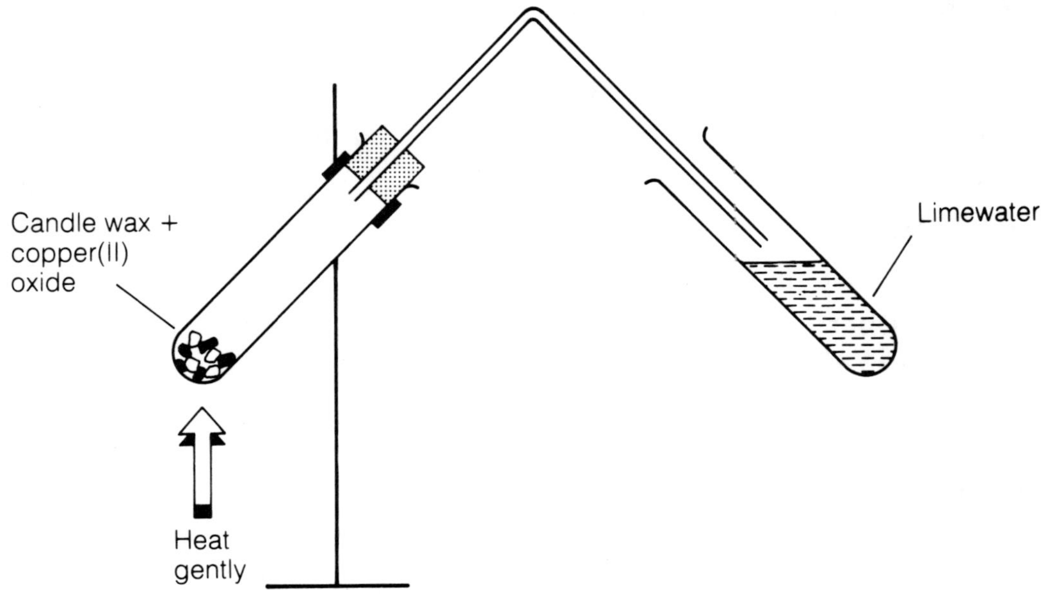

Fig. 16.2(b) Oxidation products of candle wax

TOPIC 17 Alkenes and alcohols

Experiment 17.1
Heating liquid paraffin

1 Set up the apparatus shown in Fig. 17.1. Put rocksil (or cotton wool) to a depth of 2 cm in the test-tube. Use a teat pipette to add 2–3 cm^3 of liquid paraffin. Put some broken pot in the middle of the test-tube.

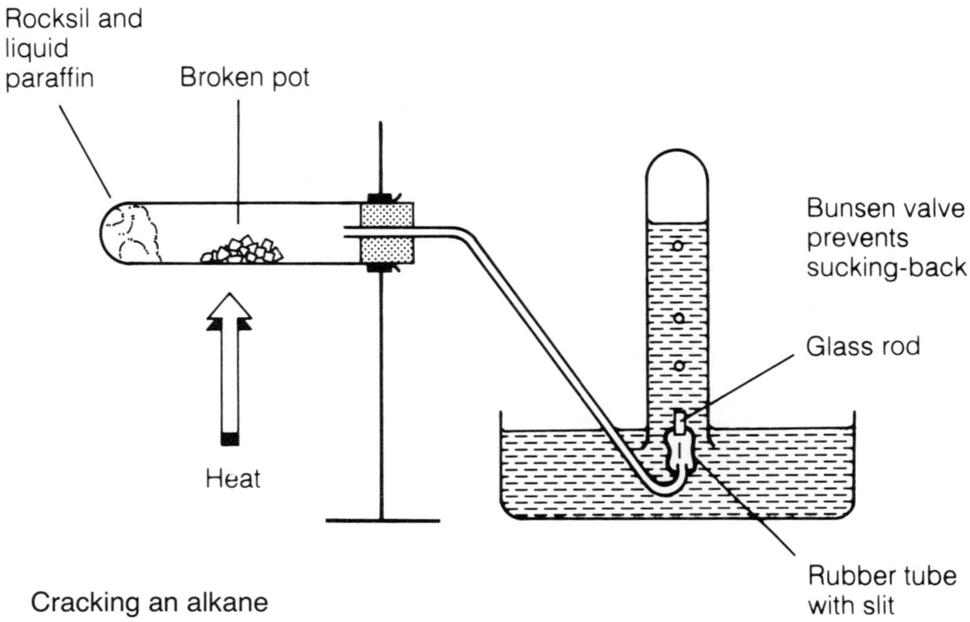

Fig. 17.1 Cracking an alkane

2 Heat the pieces of pot strongly. From time to time, warm the rocksil.

3 The first test-tube of gas you collect will be the air that has been driven out of the test-tube. Discard this, and then start to collect the gas that is evolved. When a test-tube becomes full, cork it while the open end of the test-tube is still under water. Collect three tubes, and stand them in a rack. When you stop heating, lift the delivery tube out of the trough.

4 Smell the gas. Try to describe the smell.

5 Does it burn? If it does, note the appearance of the flame.

6 To one test-tube of gas, add a few drops of a solution of bromine in trichloroethane. Note what happens. What kind of compound gives this reaction?

7 Do the tests in steps **4**, **5** and **6** on liquid paraffin.

8 Liquid paraffin is a mixture of alkanes. When it is heated in this way, it is *cracked*, that is, split up into hydrocarbons with smaller molecules. Your tests show something of the nature of the product of cracking. Write two sentences to summarise what you have found out.

Experiment 17.2
Investigating unsaturation in fats and oils

We are always reading advertisements for unsaturated fats and oils. In this experiment, you can compare the degree of unsaturation of some foods.

1 With a teat pipette, add five drops of olive oil to 5 cm^3 of trichloroethane in a conical flask.

2 Fill a burette with a dilute solution of bromine water (0.020 mol/l, see Appendix).

3 Read the burette. Run the bromine water slowly into the oil solution. Shake vigorously after each addition. The yellow colour of bromine will disappear as bromine reacts with the oil. Continue adding bromine water until a permanent yellow colour is obtained (Fig. 17.2).

4 Read the burette. By difference, find the volume of bromine water needed in the titration.

5 Repeat the experiment with 5 drops of melted margarine, 5 drops of cooking oil, 5 drops of melted butter.

6 Write a summary of your results. Which of the foods is the most unsaturated? Which is the most saturated?

7 This comparison is only approximate. What changes in the experimental method would you have to make in order to make an exact comparison between the foods?

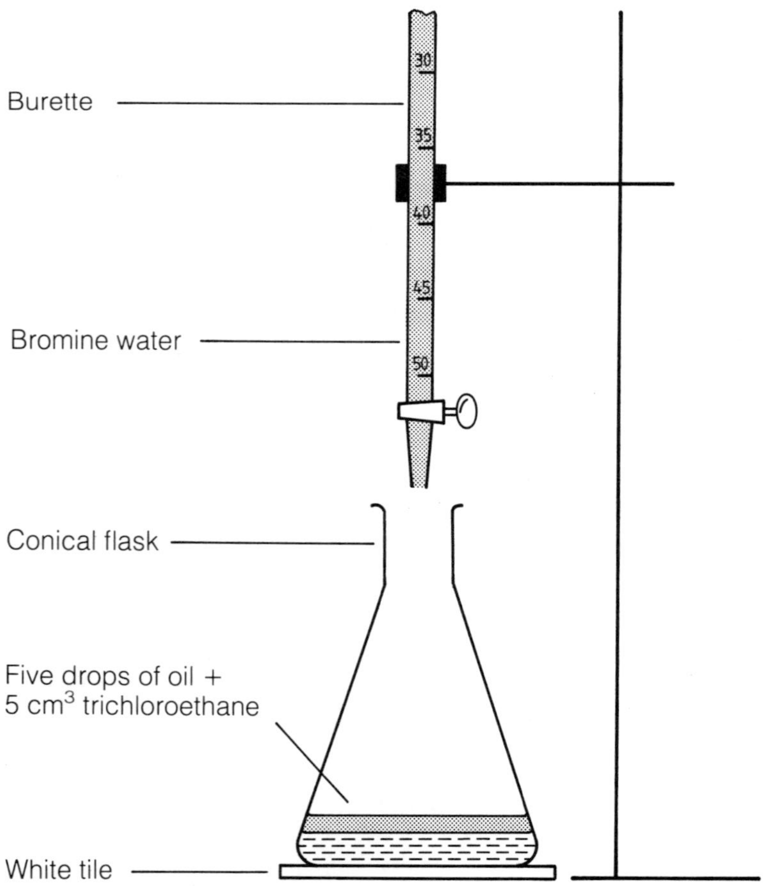

Fig. 17.2 Titrating unsaturated fats and oils

Experiment 17.3
Fermenting apple juice

1 Slice an apple. Put the slices into a 250 cm^3 beaker, and cover them with water. Boil the water, and keep it on the boil for 5 minutes.

2 When the apple and water have cooled, filter into a 250 cm^3 flask.

3 Add one spatula measure of either fresh yeast or dried yeast to the juice.

4 Set up the apparatus shown in Fig. 17.3a, and leave it for a week.

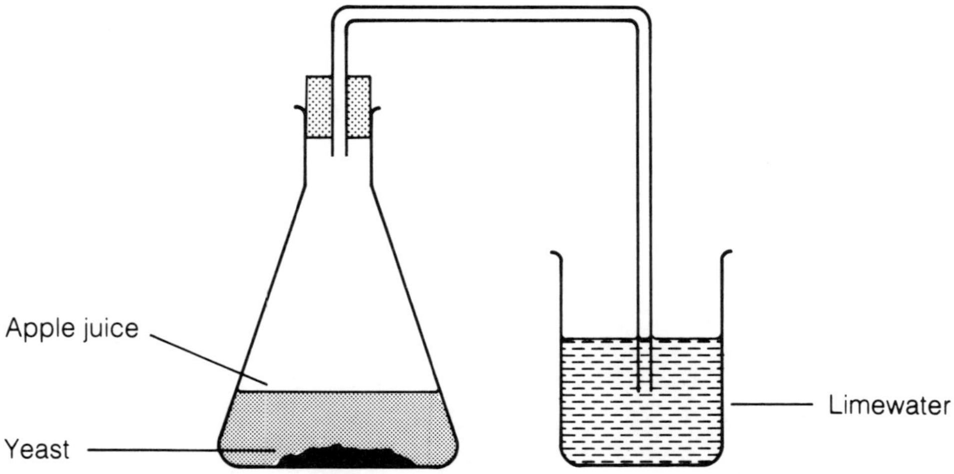

Fig. 17.3(a) Fermentation

5 After a week, decant the liquid into a round-bottomed flask. Note any change in the limewater. Assemble the apparatus shown in Fig. 17.3b.

6 Distil. Record the temperature at which liquid begins to distil over. Collect the distillate in 1 cm^3 portions.

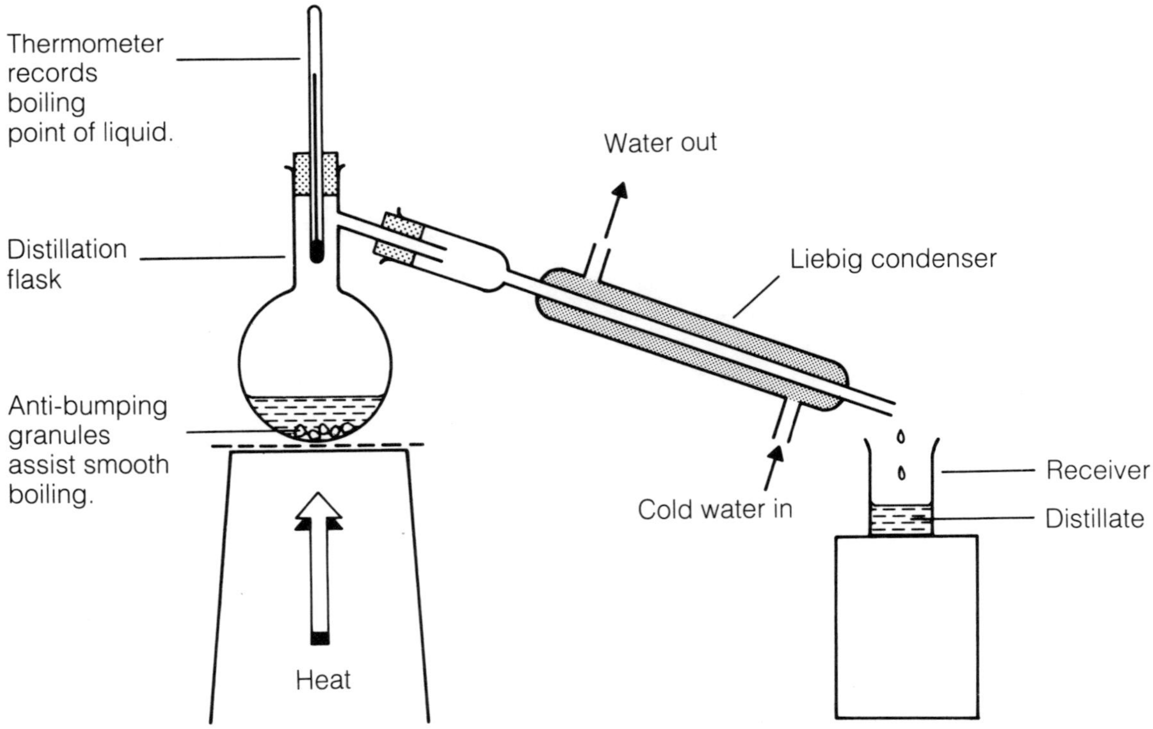

Fig. 17.3(b) Distillation

7 Pour the fractions of distillate on to watch glasses. Smell them. Attempt to ignite them with a lighted splint. Make a note of the way in which the different fractions behave.

The substance which distils at 78 °C is ethanol. It is made by the action of enzymes in yeast on sucrose, the sugar present in fruits.

sucrose → ethanol + carbon dioxide

Note: If you have other fruit, try it. If you have no fruit, use sucrose instead and add a little ammonium phosphate as a yeast nutrient.

Experiment 17.4
Studying some reactions of ethanol

Dehydration

1 Set up the apparatus in Fig. 17.4. With a teat pipette, add 5 cm^3 of ethanol to the rocksil.

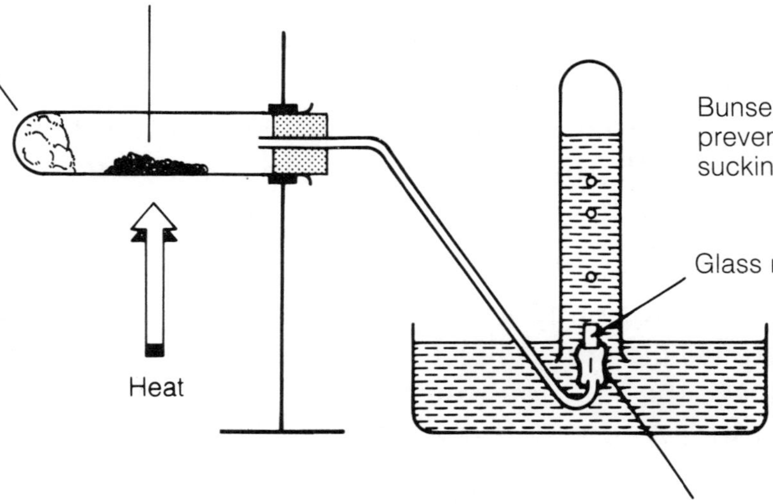

Fig. 17.4 Dehydration of ethanol

2 Heat the aluminium oxide. Collect three test-tubes of gas. Discard the first because it contains mostly air. Note the smell of the gas.
When you stop heating, lift the delivery tube out of the trough to avoid sucking-back.

3 To one test-tube of gas, apply a lighted splint.

4 To the other, add a few drops of bromine in trichloroethane.

5 Can you identify the product? It has been formed by the dehydration of ethanol:

ethanol → ———— + water
C_2H_5OH → ———— + H_2O

Oxidation

1 Take 5 cm^3 of ethanol in a boiling tube. Add 5 cm^3 of a solution of acidified sodium dichromate(VI). Warm gently. Note the change of colour. This reaction was the basis of the first breathalyser test.

2 Smell the vapour of the substance produced. Do you recognise it? Hold a piece of damp blue litmus in the vapour. Note any change.
You have oxidised ethanol to ethanoic acid.

ethanol + oxygen from oxidising agent → ethanoic acid + water
$C_2H_5OH(l) + 2[O] \rightarrow CH_3CO_2H(l) + H_2O(l)$

Experiment 17.5
Making polymers

Rayon

Rayon is a fibre made from cellulose.

1 Make 100 cm^3 of a saturated solution of copper(II) sulphate in a beaker.

2 Add sodium hydroxide solution, and stir. Copper(II) hydroxide precipitates.

3 Add 50 cm^3 of ammonia solution (0.880 ammonia) to dissolve the precipitate. **Take care; wear safety glasses.**

4 Add two torn-up filter papers (15 cm). Stir the solution thoroughly. The papers will gradually dissolve to give a viscous solution. You need to spend time on this stage to get a good result.

5 Fill a beaker with dilute hydrochloric acid.

6 Take some of the solution in a teat pipette with a long jet. Make sure that there are no particles clogging the jet. With the jet of the teat pipette below the surface, squeeze the solution into the dilute acid. Move the jet slowly through the solution. A thread of rayon will be formed.

7 Remove a length, and dry it on filter paper.

Casein

Casein is the plastic used to make knitting needles.

1 Pour off the top layer from a bottle of milk. Take 200 cm^3 of the bottom layer (skimmed milk) in a beaker. Add 20 cm^3 of bench ethanoic acid. The milk will separate into curds and whey.

2 Pour the milk through a coarse filter, such as muslin, into another beaker. The residue in the filter is the curds. It is casein.

3 Dry the casein on a filter paper. Then mould it into a shape. You could use it to line a small filter paper box or to make a ring in the bottom of a small beaker. Leave it to dry, and then cover the mould with a 10% solution of methanal (formalin) for 2 days. Pour off the formalin and let the article harden. The article will contract and be easily removed from the mould.

Urea–formaldehyde resin

This experiment needs a fume cupboard for step **2**.

1 In a beaker, take 25 g of urea and 50 g of formalin (a 40% solution of methanal). Stir.

2 Heat the mixture in a fume cupboard until it goes white and cloudy. This cloudiness is due to the formation of urea–formaldehyde resin. Divide the resin into three portions.

3 Put some cotton wool in the bottom of a plastic cup. Pour some of the resin on to it. Add a few drops of hydrochloric acid. The resin sets hard.

4 In a plastic cup, mix some of the resin with cotton wool or sawdust. Add a few drops of dilute hydrochloric acid. You will obtain a dough. Press the dough into a mould, and allow it to harden.

5 Coat one side of a microscope slide with resin. Add two drops of dilute hydrochloric acid. Place another slide on top. Hold them together with a small weight until the resin hardens.
Wearing safety glasses, hit the slide with a metal object. If you have been successful, you should have made safety glass which does not splinter when it breaks.

Nylon

1 Solution A is a 5% solution of 1,6-diaminohexane in water.

2 Solution B is a 5% solution of hexanedioyl chloride in 1,1,1-trichloroethane.

3 Place 5 cm^3 of solution B in a 50 cm^3 (or smaller) beaker. Carefully add 5 cm^3 of solution A to form a separate layer.

4 A solid forms at the junction of the two solutions (see Fig. 17.5). Pull out a fibre of the substance with forceps, and wrap it round a glass rod. As you remove the fibre, more forms at the interface. The fibre is a strand of nylon.

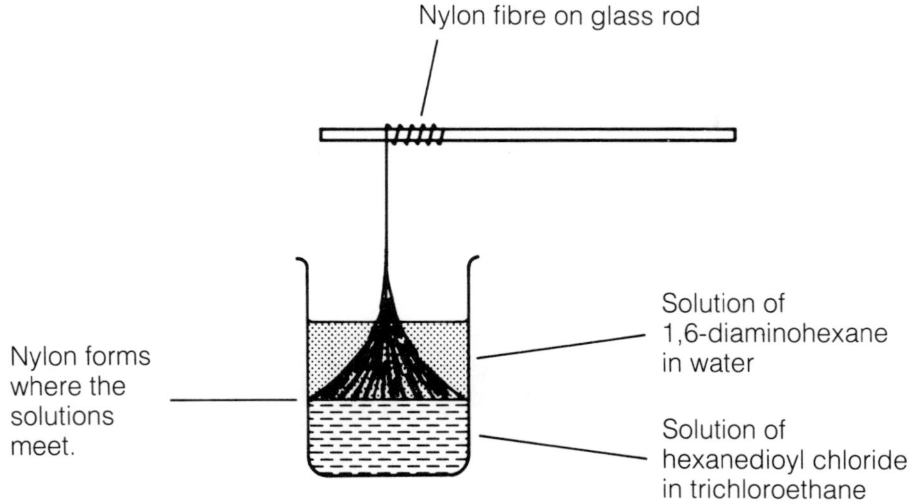

Fig. 17.5 Making nylon

Experiment 17.6
A key to identifying fabrics

Assemble a number of different fabrics. Do five tests on a sample of each fabric.

1 Heat a piece on a tin lid (Fig. 17.6a).

2 Hold a piece with tongs in a Bunsen flame (Fig. 17.6b).

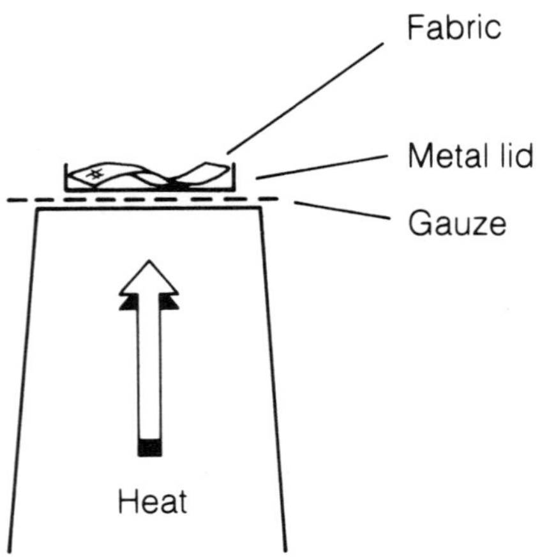

Fig. 17.6(a) Heating a piece of fabric on a tin lid

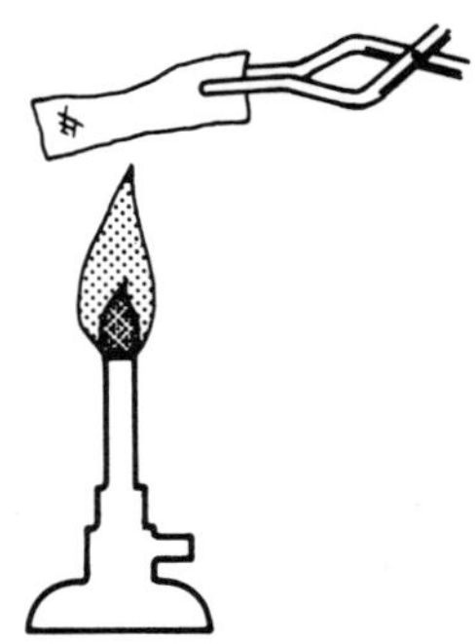

Fig. 17.6(b) Heating a piece of fabric over a Bunsen

3 Examine the burned fabric. Note the colour and the appearance of the edges.

4 Bend the burned fabric between your fingers. Does it bend or snap?

5 Pull out a single fibre of the fabric. Put it on to a microscope slide with a drop of water. Cover it with another slide. Sketch what you see under the microscope.

6 Use the key to identify the fabric.

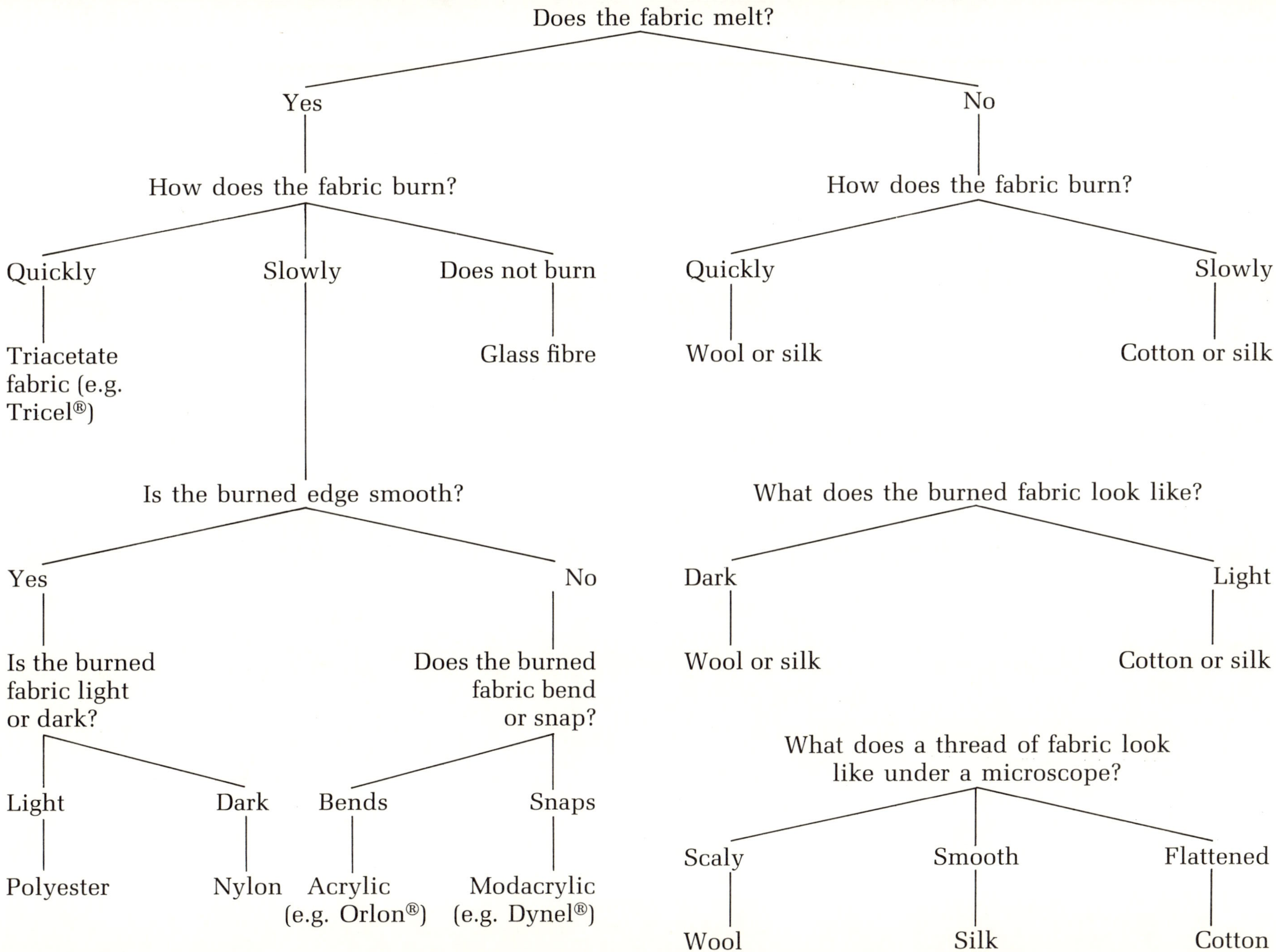

Does the fabric melt?
Yes
No
How does the fabric burn?
Quickly
Slowly
Does not burn
Triacetate fabric (e.g. Tricel®)
Glass fibre
Is the burned edge smooth?
Yes
No
Is the burned fabric light or dark?
Light
Dark
Polyester
Nylon
Does the burned fabric bend or snap?
Bends
Snaps
Acrylic (e.g. Orlon®)
Modacrylic (e.g. Dynel®)
How does the fabric burn?
Quickly
Slowly
Wool or silk
Cotton or silk
What does the burned fabric look like?
Dark
Light
Wool or silk
Cotton or silk
What does a thread of fabric look like under a microscope?
Scaly
Smooth
Flattened
Wool
Silk
Cotton

TOPIC 18 Heat changes in chemical reactions

Experiment 18.1
Some reactions which involve heat changes

1 Into a boiling tube, put 20 cm^3 of water. Put a thermometer in the water, and wait until it reads a steady temperature. Record the temperature.

2 Tip in 10 g of sherbet, and stir with a thermometer. Record the new temperature.

3 Sherbet is a mixture of citric acid and sodium hydrogencarbonate. Try the experiment again with these ingredients. Dissolve 5 g of citric acid in 25 cm^3 of water in a boiling tube. Allow it to come to room temperature, and record the temperature. Weigh 6 g of sodium hydrogencarbonate. Add it all at once with rapid stirring. Record the new temperature.

4 Repeat step **1**. Tip in 5 g of ammonium chloride, and stir. Record the new temperature.

5 Repeat steps **1** and **2**, using 5 g of potassium nitrate.

6 Put 5 g of anhydrous copper(II) sulphate into a dry boiling tube. Insert a thermometer. Add 3 cm^3 of water. Record the steady temperature attained.

7 From a measuring cylinder, pour 20 cm^3 of bench hydrochloric acid into a small beaker. Insert a thermometer. Record the temperature. In a measuring cylinder, take 20 cm^3 of bench sodium hydroxide. Pour it into the acid, and stir. Record the new temperature.

8 Copy and complete this passage.
When there is a rise in temperature during a chemical reaction, it shows that the reaction gives out ——————. Such a reaction is called a(n) —————— reaction. When there is a drop in temperature during a chemical reaction, it shows that the reaction —————— —————— ——————. Such a reaction is called a(n) —————— reaction. The reactions observed in this experiment include reactions of both types. The reactions in which the temperature rose are ————————————; these are called —————— reactions. The reactions in which the temperature fell are ————————————; these are called —————— reactions.

Extension *Neutralisation*

1 Take 100 cm^3 of 1.0 mol/l hydrochloric acid in a measuring cylinder. Record its temperature.

2 Take 100 cm^3 of 1.0 mol/l sodium hydroxide solution in a measuring cylinder. Record its temperature.

3 Obtain a clean dry 250 cm^3 beaker. Pour both solutions at the same time into the beaker. Record the highest temperature reached.

4 Repeat the experiment for all the combinations shown in the table.

Table Combinations of acids and alkalis

Acid	*Alkali*	*Temperature rise in °C*
Hydrochloric acid (1 mol/l)	Sodium hydroxide solution (1 mol/l)	
Ethanoic acid (1 mol/l)	Sodium hydroxide solution (1 mol/l)	
Ethanoic acid (1 mol/l)	Potassium hydroxide solution (1 mol/l)	
Nitric acid (1 mol/l)	Potassium hydroxide solution (1 mol/l)	
Sulphuric acid (0.5 mol/l)	Ammonia solution (1 mol/l)	
Hydrochloric acid (1 mol/l)	Ammonia solution (1 mol/l)	

5 What pattern do you observe in your results? Suggest an explanation for this pattern. (Think about what the six reactions have in common.) Write ionic equations for the six reactions.

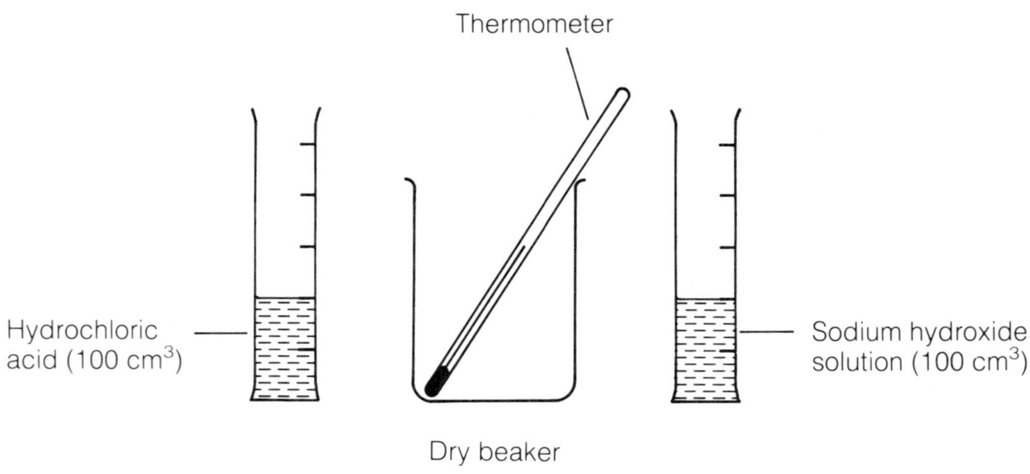

Fig. 18.1 Neutralisation

Experiment 18.2
Thermometric titration

In this experiment, sodium hydroxide solution is titrated against standard hydrochloric acid (Fig. 18.2a). The end point of the titration is found by noting the maximum temperature attained.

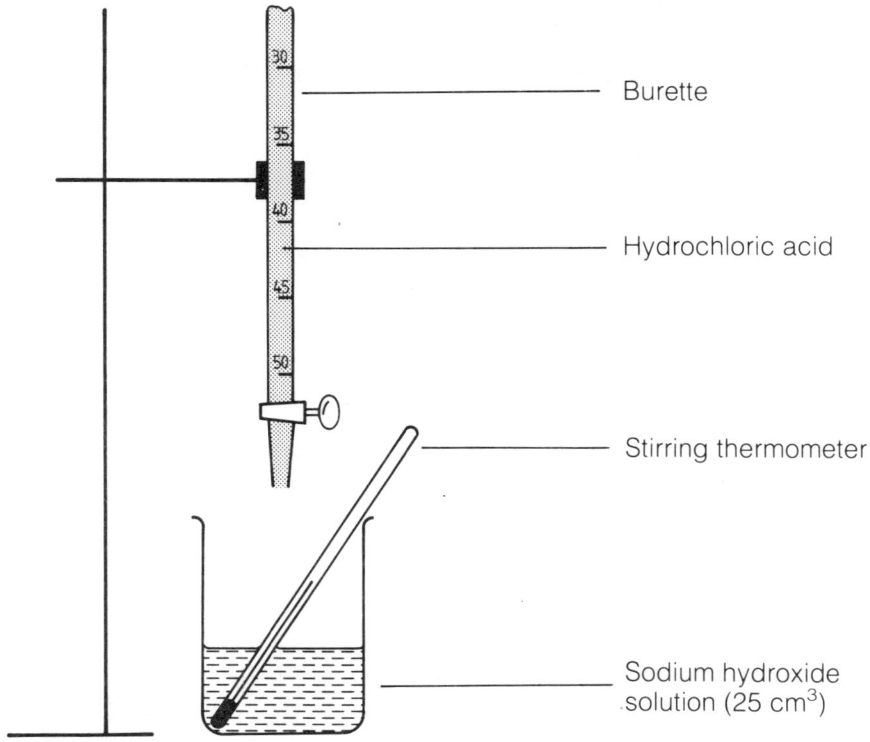

Fig. 18.2(a) Thermometric titration

1 Fill a burette with standard hydrochloric acid (1 mol/l is suitable). Place a thermometer in the bottle of hydrochloric acid, and record its temperature.

2 Measure 25 cm^3 of the sodium hydroxide solution of unknown concentration into a 100 cm^3 beaker. Place a thermometer in the solution, and record its temperature.

3 Make out a copy of the table, ready for your results.

Table Temperature during titration
Average of the temperatures of hydrochloric acid and sodium hydroxide = —°C

Volume of acid in cm^3	*Temperature in °C*	*Temperature rise in °C*
1		
2		
3		
4		
5		
6		
7		
etc.		

4 Run in 1 cm^3 of hydrochloric acid from the burette. Use the thermometer to stir the solution, and record the temperature. Enter the reading in your table.

5 Continue to add 1 cm^3 portions of hydrochloric acid, stir and record the temperature. Stop when the temperature starts to fall.

6 Subtract the average of the temperatures of the acid and alkali from the temperatures you have recorded. Enter the values you obtain for the temperature rise in the table.

7 Plot the temperature rise against the volume of acid added (Fig. 18.2b).

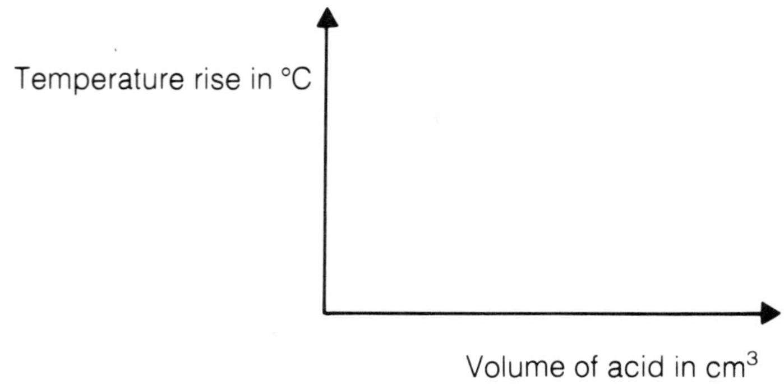

Fig. 18.2(b) Plotting temperature rise against volume of acid

From the graph, read the maximum temperature rise.
Read the volume V of acid which produces this rise. This is the volume of acid which neutralises 25 cm^3 of sodium hydroxide solution.
You can now calculate the concentration c of the sodium hydroxide solution.

Amount of HCl = Volume × Concentration = $V \times 10^{-3} \times 1.0$ mol

Amount of NaOH = Volume × Concentration = $25 \times 10^{-3} \times c$ mol

$$c = \frac{V \times 1.0}{25} \text{ mol/l}$$

Heat is lost to the surroundings from the beaker and the solutions. Does this affect the accuracy of your result?
Should you do something to reduce the heat loss? If you think so, say what change or changes you would make.

Experiment 18.3
Observing temperature changes when copper is displaced from copper(II) sulphate solution

The reaction to be studied is

zinc + copper(II) sulphate solution → copper + zinc sulphate solution

$Zn(s) + CuSO_4(aq) \rightarrow Cu(s) + ZnSO_4(aq)$

1 With a measuring cylinder, take 25 cm^3 of copper(II) sulphate solution. Pour the solution into a polystyrene cup.

2 Weigh about 3 g of zinc powder on a piece of paper.

3 Put the thermometer through the hole in the lid. Record the temperature every 30 seconds for 2.5 minutes (see Fig. 18.3a).

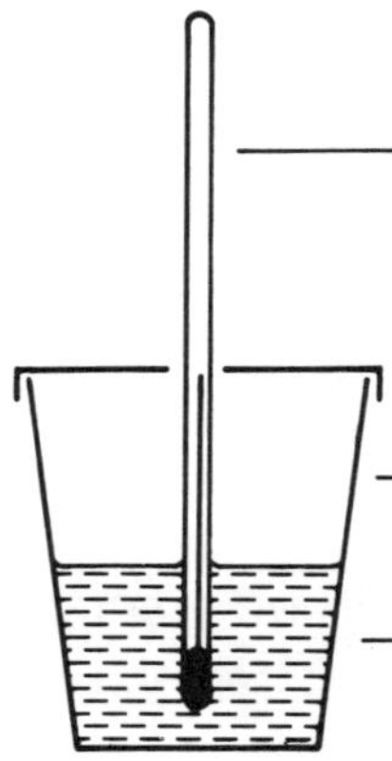

Fig. 18.3(a) Recording the temperature

4 At 3 minutes, add the zinc powder.

5 Stir, and record the temperature every 30 seconds for 6 minutes. Copy the table, and enter your results.

Table Temperature changes when copper is displaced from copper(II) sulphate solution

Time in min	0.0	0.5	1.0	1.5	2.0	2.5	3.0	3.5	4.0	4.5
Temperature in °C										
Time in min	5.0	5.5	6.0	6.5	7.0	7.5	8.0	8.5	9.0	
Temperature in °C										

6 Plot the temperature against the time (see Fig. 18.3b).

7 Find the maximum temperature rise. Read the maximum temperature from your graph. Read the steady temperature at 0 to 2.5 minutes. Subtract to obtain the temperature rise.

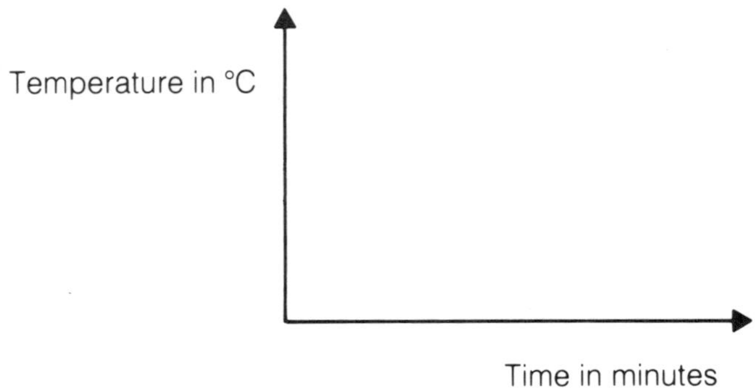

Fig. 18.3(b) Plotting the results

Extension

There is a more accurate way of finding the maximum temperature rise. Use the graph you plotted in step **6**.

- **a** Draw a curve through your points and extrapolate it (continue it back) to 3.0 minutes.
- **b** Read the temperature at 3.0 minutes from the curve.
- **c** Subtract the steady temperature at 0 to 2.5 minutes from this maximum temperature.
- **d** Why is this maximum temperature read from the graph a better measure of the rise in temperature than the temperature recorded by the thermometer?
- **e** What amount of copper(II) sulphate have you used? What fraction of 1 mol is this? What would be the temperature rise if you used 1 mol of copper(II) sulphate and an excess of zinc?

8 Repeat the experiment, using 3 g of iron filings instead of zinc.

9 Repeat the experiment, using 3 g of magnesium powder.

10 Tabulate your results. Rank the metals in order of the heat generated in the reactions. Can you explain why they should fall in this order?

Experiment 18.4
Comparing different fuels

The heat given out when a certain mass of fuel burns is used to raise the temperature of a known mass of water. The rise in temperature produced by different fuels can be compared with the mass of fuel burned. Work in which a rise in temperature is measured and used to assess a heat output is called *calorimetry*.

1 The apparatus is shown in Fig. 18.4. Weigh the spirit lamp about half full of ethanol to give m_1.

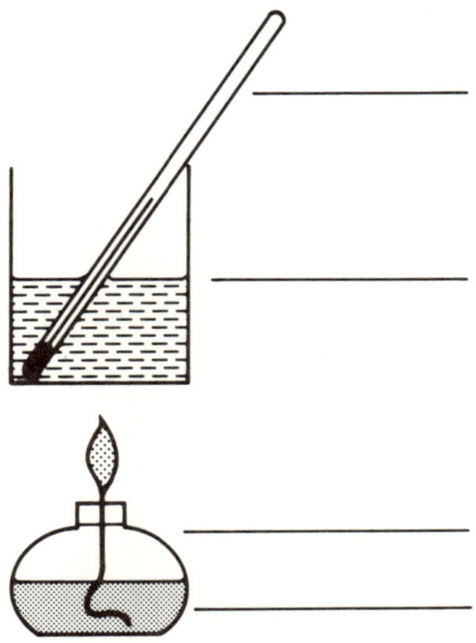

Fig. 18.4 Apparatus for calorimetry

2 Take the temperature T_1 of the water in the can, and record it.

3 Place the lamp under the can, and light it. Stir the water with the thermometer. When the temperature has risen by about 20 °C, extinguish the lamp. Record the final temperature T_2.

4 Reweigh the spirit lamp and its contents to give m_2.

5 The mass of fuel burned $= m_1 - m_2$
The rise in temperature $= T_2 - T_1$

Rise in temperature produced by burning 100 g of fuel $= \dfrac{T_2 - T_1}{m_1 - m_2} \times 100$

6 Repeat the whole of the procedure for another fuel. You can use methanol, propanol, butanol, kerosene (but **not** petrol) and hexane. In each case, find the rise in temperature that results from burning 100 g of the fuel. The same calorimeter must be used, and the mass of water must always be the same.

7 Rank the fuels you have tested in order of *calorific value*.

8 Of the heat produced by the fuel, some is lost to the surroundings. How badly will this affect the comparison of the different fuels? What could you do to reduce the heat lost in this way?

9 Repeat one of the measurements you have made, employing your idea for reducing heat loss. What is the temperature rise this time?

TOPIC 19 Speeds of reactions

Experiment 19.1
Does the speed at which a solid reacts depend on the size of the solid particles?

Calcium carbonate occurs naturally as chalk, as limestone and as marble. When calcium carbonate reacts with dilute hydrochloric acid, the gas carbon dioxide is given off. You can use large marble chips, small marble chips and medium-sized marble chips to see which size of particles react the fastest.

1 Using a top-loading balance, find two marble chips which together weigh about 10 g. Then find five to six chips which together weigh 10 g. Lastly, weigh out 10 g of the smallest chips you can find.

2 Take 15 cm^3 of dilute hydrochloric acid in a measuring cylinder. Pour it into the side-arm tube shown in Fig. 19.1. Make sure the delivery tube ends beneath the surface of the water in the beaker.

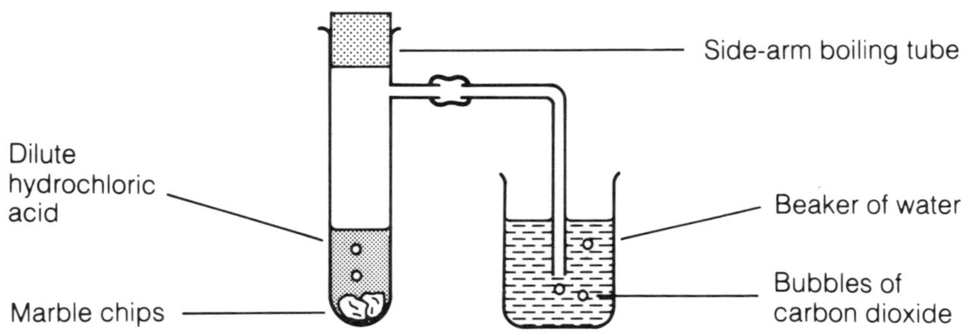

Fig. 19.1 Apparatus for counting bubbles

3 Take a rubber bung in one hand and the large marble chips in the other. Drop the large chips into the acid. Quickly put the bung into the top of the side-arm tube. Make sure it fits tightly. At the same time, start a stopclock. Count the number of bubbles given off in 1 minute.

4 Wash out the apparatus. Repeat the experiment, using the five to six medium-sized chips.

5 Repeat, using the 10 g of small chips.

6 Copy and complete the following.

Number of bubbles given off in 1 minute by 10 g of large chips = ————

Number of bubbles given off in 1 minute by 10 g of medium chips = ————

Number of bubbles given off in 1 minute by 10 g of small chips = ————

The chips which gave off most bubbles in 1 minute were the ———— chips.

The chips which reacted fastest were the ———— chips.

Try to think of an explanation for your results.

Experiment 19.2
How does the concentration of a reactant affect the time taken for the completion of a chemical reaction?

Pour 5 cm^3 of dilute sulphuric acid into a test-tube. Add a small piece of magnesium ribbon. Note the reaction. Test the gas formed with a lighted splint. What happens? What is the gas? Copy and complete the word equation

magnesium + sulphuric acid → ———— (g) + ———— (aq)

This is the reaction to be studied. The method to be used is as follows.

1 With a measuring cylinder, take 20 cm^3 of bench sulphuric acid. Pour it into a 50 cm^3 beaker.

2 Drop a 3 cm length of magnesium ribbon into the beaker. As soon as it comes into contact with the acid, start a stopclock. Count the number of seconds until the piece of magnesium has completely disappeared.

3 Copy the table shown below. Enter the time in the table.

4 Wash out the beaker. Repeat the experiment, using 17.5 cm^3 of acid and 2.5 cm^3 of water. Use exactly the same length of magnesium ribbon.

5 Work through the table. Use all the different mixtures of acid and water listed. Note the time for complete reaction in each case.

Table Times taken for complete reaction

Volume of acid in cm^3	*Volume of water in cm^3*	*Time taken for reaction in seconds*
20.0	0	
17.5	2.5	
15.0	5.0	
12.5	7.5	
10.0	10.0	
7.5	12.5	
5.0	15.0	
2.5	17.5	

6 On graph paper, plot the volume of acid used against the time taken for the magnesium to react completely and disappear (Fig. 19.2).

7 Write the answers to the following questions.
In which reaction did the magnesium disappear fastest?
In which reaction did the magnesium disappear most slowly?
In which reaction was the acid most concentrated?
In which reaction was the acid most dilute?
What happened to the speed of the reaction when you diluted the acid by replacing some of the acid by water?

Write two sentences to summarise what you have found out. What happens to the speed of this reaction when you alter the concentration of one of the reactants?

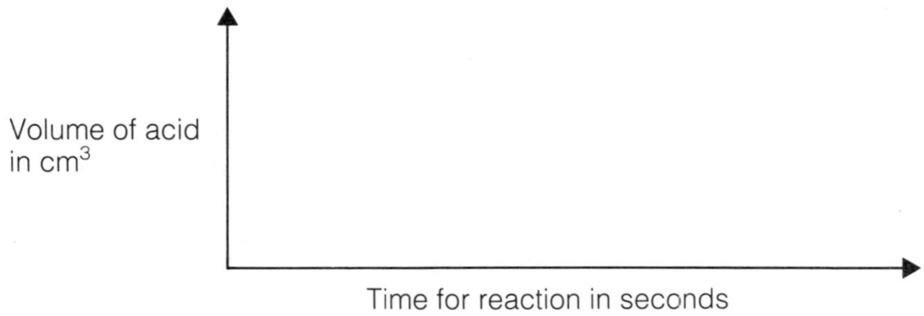

Fig. 19.2 Plotting the volume of acid against the time for reaction

Experiment 19.3
Investigating the effect of concentration on the speed of the reaction between acid and thiosulphate

Note: Sodium thiosulphate has the formula $Na_2S_2O_3$. It resembles sodium sulphate, Na_2SO_4, with one less oxygen atom and one more sulphur atom. The additional sulphur atoms can be driven out by an acid.

sodium thiosulphate	+	hydrochloric acid	→	sulphur	+	sulphur dioxide	+	sodium chloride
$Na_2S_2O_3(aq)$	+	$2HCl(aq)$	→	$S(s)$	+	$SO_2(g)$	+	$2NaCl(aq)$

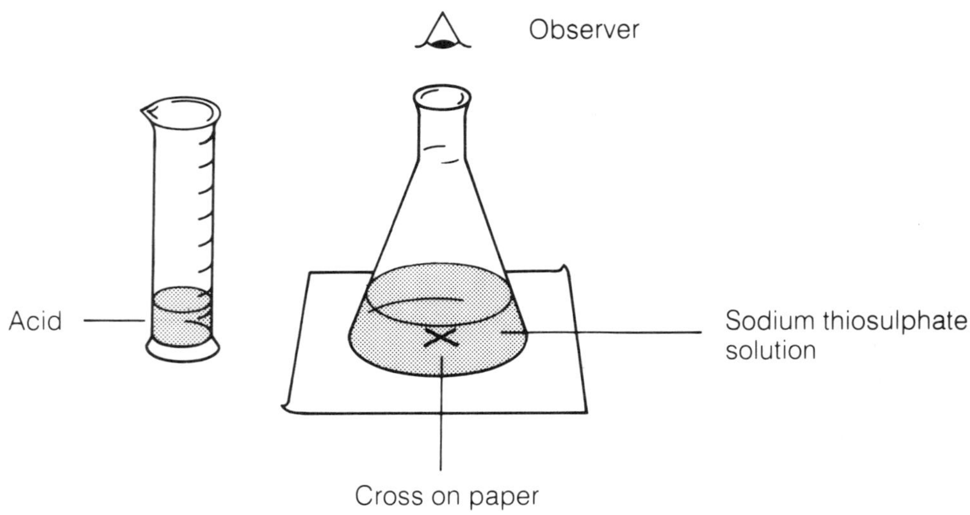

Fig. 19.3 Apparatus for studying the acid–thiosulphate reaction

This reaction can be studied as follows.

1 Make a copy of the table, ready for your results.

2 Measure out 50 cm^3 of sodium thiosulphate solution in a measuring cylinder. (*Note*: 40 g/l.) Pour it into a conical flask. Keep this measuring cylinder for thiosulphate only. It is important that no acid gets into it.

3 On a piece of paper, write a cross. Stand the conical flask over this cross (see Fig. 19.3).

4 Measure 5 cm^3 of bench hydrochloric acid from a measuring cylinder. Keep this measuring cylinder for acid only.

5 Pour the acid into the sodium thiosulphate solution. Swirl. At the same time, start a stopwatch.

6 Watch the cross. Time how long it takes for the deposit of sulphur to hide the cross from sight. Enter this time into the results table.

7 Wash out the conical flask. Repeat the experiment. This time, use 40 cm^3 sodium thiosulphate solution and 10 cm^3 of water. Keep everything else the same as before.

Table Time taken to hide the cross from view

Volume of sodium thiosulphate solution in cm^3	*Volume of water in cm^3*	*Time taken for cross to vanish in seconds*
50	0	
40	10	
30	20	
20	30	
10	40	

8 Answer these questions.
In which experiment did the cross disappear fastest?
In which experiment did the cross take longest to disappear?
Which reaction was the fastest? And which was the slowest?
Which solution was the most concentrated? And which was the most dilute?
What happens to the speed of this chemical reaction when we alter the concentration of one of the reactants?

9 How could you find out what happens when you alter the concentration of hydrochloric acid? Say what solutions you would use. List the experiments you would do. Say what you would do with your results.

Experiment 19.4
Investigating the effect of temperature on the speed of a chemical reaction

1 Measure in a measuring cylinder 50 cm^3 of sodium thiosulphate solution. (*Note*: 8 g/l). Pour this solution into a conical flask.

2 Do the first experiment at room temperature. Follow the instructions for Experiment 19.3. Note the temperature of the reacting solution.

3 For the second experiment, measure a second 50 cm^3 of sodium thiosulphate solution. Pour it into a clean flask. Warm the flask to 30 °C, as shown in Fig. 19.4. The temperature need not be *exactly* 30 °C.

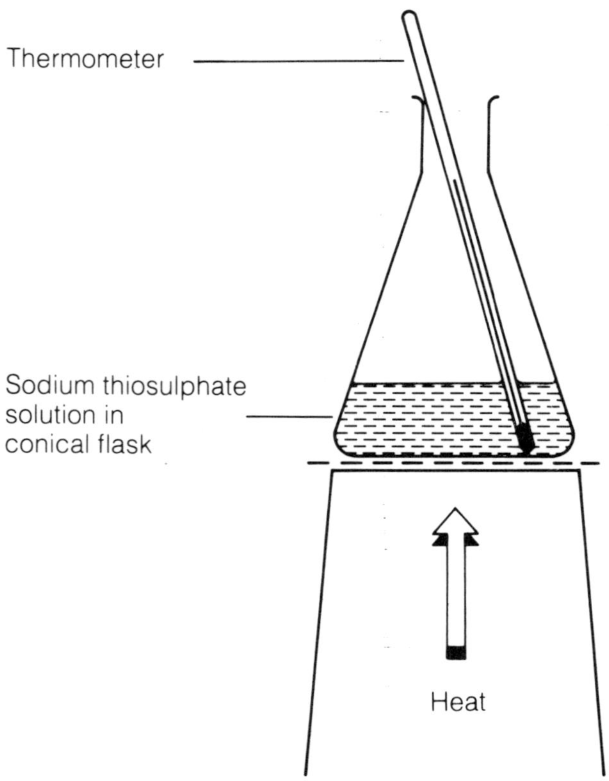

Fig. 19.4 Heating a sodium thiosulphate solution

Remove the flask from the tripod. Stand it on the cross on the piece of paper. Now, you must do four things at the same time (as nearly as you can).

a Add 5 cm^3 of hydrochloric acid.
b Swirl the flask.
c Start a stopclock.
d Read the temperature of the solution. (The temperature need not be exactly 30 °C, but you must note what the *exact* temperature is.)

Time the disappearance of the cross.

4 Repeat the experiment at 40 °C.

5 Repeat the experiment at 50 °C.

6 Repeat the experiment at 60 °C. (In each case, note what the exact temperature is.)

7 Make a table of your results.

Table Results

Temperature in °C	*Time for cross to vanish in seconds*

8 At which temperature was the time taken for the cross to vanish longest?
At which temperature did the cross disappear most slowly?
At which temperature did the cross vanish the fastest?
Which reaction was the slowest?

9 Plot your results on graph paper, as shown in Fig. 19.4.

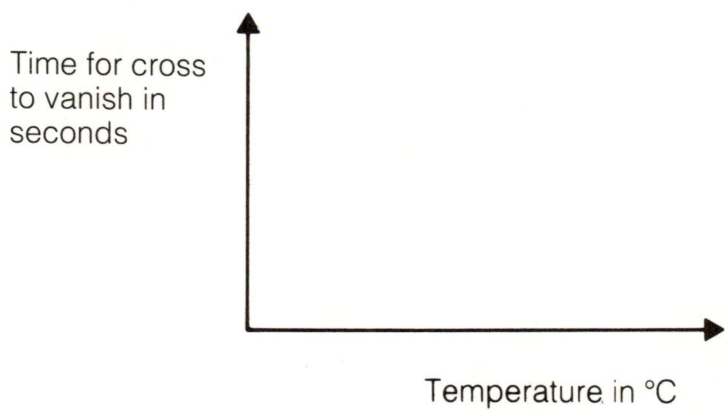

Fig. 19.4 Plotting time taken for cross to vanish against temperature

10 Write a statement of what you have found out about the effect of temperature on the rate at which this reaction takes place.

Experiment 19.5
The effect of light on a chemical reaction

1 Mix 20 cm^3 of silver nitrate solution and 20 cm^3 of potassium bromide solution. Stir the mixture.

2 Filter off the precipitate of silver bromide. Wash the precipitate with distilled water.

3 Spread the precipitate on a filter paper. Cover half of it with a piece of card.

4 Either put the precipitate on a window ledge for 15 minutes or direct a beam of light on to it for 5 minutes.

5 Remove the card. Compare the silver bromide that was exposed to light with the silver bromide that was covered. Which is the darker? Darkening of the precipitate is due to the formation of silver.
Does the formation of silver from silver bromide take place faster in the presence of light or the absence of light?

6 Decide what solutions you need to mix to make silver chloride. Mix the solutions, and repeat steps **2–5** with silver chloride.

7 Repeat the experiment with silver iodide.

8 Write a few sentences to say what you have found out about
- **a** the effect of light on the formation of silver,
- **b** the difference between silver bromide, silver chloride and silver iodide.

Experiment 19.6
Catalysis

Note: Hydrogen peroxide is a colourless odourless liquid. It has the formula H_2O_2. It is sold as an aqueous solution. People use it as a bleach to make hair blonde and also as an antiseptic. It is kept in brown bottles because it decomposes slowly to give oxygen and water:

hydrogen peroxide → oxygen + water

$$2H_2O_2(aq) \rightarrow O_2(g) + 2H_2O(l)$$

The decomposition is slow at room temperature. In this experiment, you can try various substances to see whether they speed up the evolution of oxygen.

1 Take 5 cm^3 of hydrogen peroxide solution (2 volume) in a test-tube. Test for oxygen, as in Fig. 19.6. Does the splint light?

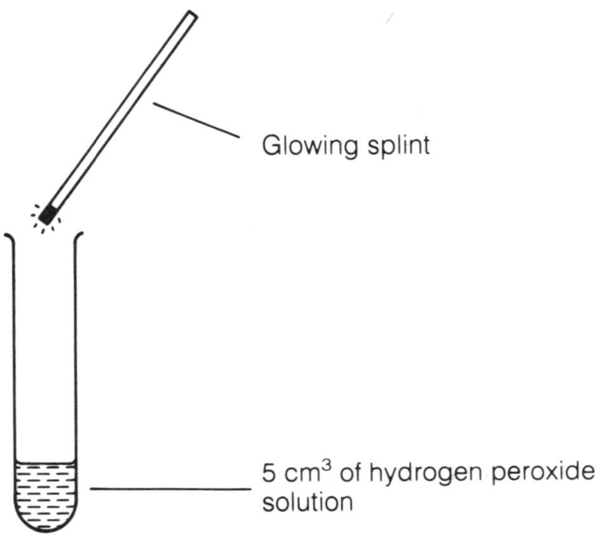

Fig. 19.6 Testing for oxygen

2 Add a few grains of copper(II) oxide, CuO, and try again.

3 Take a fresh 5 cm^3 of hydrogen peroxide solution. Add a few grains of manganese(IV) oxide, MnO_2, and try again.

4 Repeat the experiment with powdered manganese(IV) oxide.

5 Repeat the experiment with fresh hydrogen peroxide and a small cube of potato.

6 Repeat the experiment, this time adding a small cube of liver.

7 In which experiment was the evolution of oxygen fastest?
In which experiment was the evolution of oxygen slowest?
Explain the difference you found between step **3** and step **4**.
A substance which speeds up a chemical reaction, *without being used up in the reaction*, is called a *catalyst*. Manganese(IV) oxide catalyses the decomposition of hydrogen peroxide. The catalyst is not used up. At the end of the reaction, it is as good as new. You can filter off the manganese(IV) oxide which you used in step **3**, and use it in another experiment. Try this if you have time.

Appendix
List of chemicals and apparatus needed for each experiment

Safety glasses are needed for all the experiments.

Experiment 1.1
Ethanol or methylated spirit, glycerol, thermometer held in a cork, rubber band, cotton wool, clamp and stand, clock.

Experiment 1.2
Aluminium potassium sulphate, two 250 cm^3 beakers, glass rod, water bath, cotton.

Experiment 1.3
Potassium chloride, boiling tube, 250 cm^3 beaker, tripod and gauze, stirring thermometer, graph paper, potassium chlorate(V), sodium chloride, copper(II) sulphate, sodium nitrate.

Experiment 1.4
A selection of solvents, e.g. soap solution, detergent solution, ammonia, ethanol, methanol, propanone, ethyl ethanoate, 1,1,1-trichloroethane, dichloromethane, stoppered bottles, labels.

Experiment 2.1
Rock salt, pestle, mortar, beaker, glass rod, filter funnel, filter paper, tripod and gauze, evaporating basin, water bath.

Experiment 2.2
Sodium chloride, lead(II) chloride, sand, a mixture of sodium chloride, lead(II) chloride and sand, boiling tubes.

Experiment 2.3
Side-arm boiling tube, thermometer, delivery tube, test-tubes, clamp and stand, black ink.

Experiment 2.4
Distillation flask, condenser, thermometer, receiver.

Experiment 2.5
Black ink, chromatography paper, stapler, glass rod, 250 cm^3 beaker, teat pipette, oven.

Extension
Felt-tip pens, screened methyl orange. Smarties®.

Experiment 2.6
Glass tubing, chromatography paper, stapler, 250 cm^3 beaker, screened methyl orange, bromophenol blue, methylene blue, xylene cyanol.

Experiment 2.7
Blue, red or purple flower petals, pestle and mortar, propanone, teat pipette, mixture of propanone (3 volumes) and water (1 volume), boiling tube with bung, drawing pin, dilute hydrochloric acid.

Experiment 2.8
Cotton wool, test-tubes and rack, clamp and stand, glass rod, beaker, food colourings, screened methyl orange. The holes in the test-tube are made by corking a dry test-tube and heating the end of the test-tube strongly in one spot. The pressure of air inside the test-tube increases and blows a hole at the spot being heated.

Experiment 2.9
Octadecanoic acid (stearic acid), boiling tube, beaker, stirring thermometer, clock, clamp and stand.

Experiment 3.1
Pyrex test-tube, clamp and stand, iron wool, iodine.

Experiment 3.2
Tin, iodine, dichloromethane (or trichloroethane), boiling tube, beaker, crystallising dish.

Experiment 3.3
Pyrex test-tube, clamp and stand, copper strip, sulphur, balance.

Experiment 3.4
Iron filings, sulphur, magnet, clamp and stand.

Experiment 4.1
Graph paper

Experiment 5.1
6 V battery or transformer, bulb in holder, electrical wires ending in crocodile clips, metals (aluminium, copper, iron, lead, tin and zinc), lumps of sulphur, graphite and charcoal, large crystal of copper(II) sulphate-5-water, copper(II) chloride, potassium iodide, lead(II) bromide, sodium chloride, two carbon electrodes, 100 cm^3 beaker with card to fit the top.

Experiment 5.2
See Fig. 5.2 or use a commercial apparatus, solutions of copper(II) chloride, copper(II) sulphate, sodium chloride, potassium iodide, sulphuric acid.

Experiment 5.3
Microscope slide, battery or transformer, electrical leads with crocodile clips, filter paper or chromatography paper.

Experiment 5.4
Zinc dust, iodine, beaker, battery or transformer, carbon rod as anode, copper strip as cathode (makes the change at the cathode clearly visible, but a second carbon rod would do).

Experiment 5.5
Two strips of copper foil, sandpaper, battery or transformer, beaker, copper(II) sulphate solution.

Extension
Balance.

Experiment 5.6
Beakers, cardboard, specimen bottles, corks, pieces of nichrome wire, pencil leads (or carbon rods), 9 V battery or transformer, electrical wire, crocodile clips, pieces of wood, stapler, solutions of sodium sulphate, sodium hydrogencarbonate, ethanoic acid, propan-1-ol and glucose. (Light bulbs are *not* supplied. The idea is to assess the rate of evolution of hydrogen bubbles. It is expected that pupils will think of putting the solution into a specimen tube and inserting pencil leads through a cork or putting the solution into a beaker and inserting electrodes through a piece of carboard.)

Experiment 5.7
Iron key or nail, copper(II) sulphate, dilute ammonia, sandpaper, steel wool, dilute hydrochloric acid, dilute sodium hydroxide, beaker, copper wire, 4.5 V battery or transformer, strip of copper, ethanol or methylated spirit.

Experiment 5.8
Two sheets of lead (2 cm × 8 cm), 4.5 V battery, 1.25 V bulb in holder, two electrical wires with crocodile clips, 100 cm^3 beaker, piece of wood, dilute sulphuric acid.

Experiment 5.9
9 V battery or transformer, sodium hydroxide solution, dilute nitric acid, aluminium foil, detergent solution, electrical wires with crocodile clips, Congo red (or eosin or alizarin).

Extension
Copper, zinc, zinc sulphate, boric acid, dilute sulphuric acid.

Experiment 5.10
Microscope slide, filter paper or chromatography paper, electrical leads with crocodile clips, pencil lead (or carbon rod), 6 V battery or transformer, sodium chloride, red litmus, blue litmus, phenolphthalein.

Extension
Bromocresol green, methyl orange, screened methyl orange, red litmus paper, battery.

Experiment 6.1
Card, scissors, crayons or felt-tip pens.

Experiment 7.1
Test-tubes and rack, hydrochloric acid, sulphuric acid, sodium hydroxide solution, litmus, methyl orange, teat pipette, magnesium, iron, copper, zinc, lead, copper(II) oxide, sodium carbonate, limewater, splints.

Experiment 7.2
Sodium hydroxide, ammonia, calcium hydroxide, calcium oxide, test-tubes and rack, litmus, solutions of copper(II) sulphate, iron(II) sulphate, iron(III) chloride, magnesium sulphate, olive oil, 20% sodium hydroxide solution, saturated sodium chloride solution.

Extension
Solutions of aluminium sulphate, lead(II) nitrate and zinc sulphate.

Experiment 7.3
Vinegar or ethanoic acid solution, garden lime or calcium hydroxide, filter funnel and paper, evaporating basin, water bath, test-tube.

Experiment 7.4
Indicators (litmus, screened methyl orange, phenolpthalein, universal indicator), selection of acids and alkalis, a 3.0 g/l solution of ethanoic acid.

Experiment 7.5
Sulphuric acid, zinc, beaker, tripod and gauze, filter funnel and paper, water bath.

Experiment 7.6
Copper(II) oxide, sulphuric acid, beaker, spatula, glass rod, tripod and gauze, litmus paper, evaporating basin, water bath.

Experiment 7.7
Copper(II) sulphate-5-water crystals (from Experiment 7.6), side-arm boiling tube, thermometer, delivery tube, test-tube in rack.

Experiment 7.8
Solutions of sodium hydroxide and sulphuric acid, burette, pipette, pipette-filler, conical flask, screened methyl orange, evaporating basin, water bath.

Experiment 7.9
Two measuring cylinders, beaker, filter funnel and paper, watch glass, solutions of lead(II) nitrate and potassium chromate(VI), potassium iodide, iron(II) sulphate, tea, water glass, crystals or salts listed in Experiment 7.9, A chemical garden, step **3**.

Experiment 7.10
Selection of ant-acid indigestion tablets, methyl orange or bromophenol blue (must be an indicator which changes at about pH 5), bench hydrochloric acid, spatula, conical flask.

Experiment 7.11
Milk, sodium hydrogencarbonate.

Experiment 8.1
Cylinder of oxygen or generator shown in Fig. 8.1, four boiling tubes with corks, trough, combustion spoon, iron wool, magnesium ribbon, charcoal, sulphur. (**Teacher demonstration only**: sodium calcium, phosphorus.)

Experiment 8.2
Nine clean iron nails, nine test-tubes in rack, oil, anhydrous calcium chloride, soda–lime, grease, oil paint.

Extension
Petri dish, iron wool, iron nails, beaker, test-tube, balance.

Experiment 8.3
Graduated tube, iron wool, bucket or some other deep container.

Experiment 8.4
Packet of cress seeds, empty yoghurt cartons, aluminium foil, empty margarine tubs, crystallising dishes, beakers, plastic bags with ties, labels, plastic food boxes, sand, soil, cotton wool, sodium metabisulphite.

Experiment 8.5
Grass, green leaves, flower petals, 2 volume hydrogen peroxide, rack of test-tubes, sodium hydroxide, hydrochloric acid, splint, blue litmus paper, solutions of potassium iodide and potassium bromide, sulphuric acid, hexane or trichloroethane.

Extension
Gallic acid (3,4,5-trihydroxybenzoic acid), iron(II) sulphate.

Experiment 9.1
Soap solution (20 g of soap flakes dissolved in 20 cm^3 of ethanol and 100 cm^3 of warm distilled water, and made up to 1 litre), teat pipette or graduated 1 cm^3 pipette, rack of test-tubes, solutions of salts listed in the table, see Experiment 9.1.

Experiment 9.2
Burette, soap solution, conical flask, washing soda, washing powder, distilled water.

Extension
Ion exchange resin, e.g. Permutit®, lengths of glass tubing of width ½ cm, 1 cm, 2 cm and 4 cm, rubber bungs, corks and borer, rubber tubing, glass wool, cotton wool, screw clips, solutions of cobalt(II) sulphate and sodium chloride.

Experiment 9.3
Plastic cups, oil, sodium chloride solution, liquid detergent, plaster of Paris, sawdust, chips of polystyrene foam, Polyclens® etc., teat pipette.

Experiment 9.4
Powdered charcoal, soap solution, detergent solution, piece of cloth, coloured oil, magnesium sulphate, grease, solid detergent, liquid detergent.

Experiment 9.5
Selection of detergents, including a soap powder, e.g. Lux®, ethanol, universal indicator, starch–iodide paper, dilute hydrochloric acid, metal dish or a square of aluminium foil which can be made into a dish.

Experiment 9.6
Enzyme detergent, non-enzyme detergent, white cotton fabric, staining substances, e.g. coffee, tea, fruit juice, egg, beer, grass, blood (from the butcher).

Extension
Thermometer, sodium chloride.

Experiment 9.8
See list in Experiment 9.8.

Experiment 9.9
Soil samples (clay, loamy soil, chalky soil, sandy soil), filter funnels and paper, beakers, measuring cylinders.

Experiment 10.1
Test-tubes and rack, dilute hydrochloric acid, zinc, calcium, magnesium, iron, tin, aluminium, copper, lead.
(**Teacher demonstration only**: sodium and potassium.)

Experiment 10.2
Calcium, beaker, test-tube, straw, filter funnel and paper, litmus.

Experiment 10.3
Magnesium, rocksil, Pyrex test-tube, clamp and stand, delivery tube, red and blue litmus.

Experiment 10.4
Rocksil, Pyrex test-tube, iron wool, delivery tube, test-tube, trough, zinc.

Extension
Copper(II) oxide.

Experiment 10.5
Magnesium, tin, zinc, copper, solutions of copper(II) sulphate, zinc nitrate, tin(II) chloride, iron(II) sulphate.

Extension
Solutions of lead(II) ethanoate and silver nitrate.

Experiment 10.6
Sodium carbonate crystals, aluminium, tarnished silver.

Experiment 10.7
Carbon, metal oxides, e.g. aluminium oxide, calcium oxide, copper(II) oxide, iron(III) oxide, lead(II) oxide, zinc oxide, ceramic paper, tripod and gauze.

Experiment 10.8
Strips of zinc, iron, copper and a 'mystery' metal, sodium chloride solution, galvanometer or multimeter, connecting wires.

Experiment 10.9
Three nails, pliers, heat-proof mat, beaker, emery paper.

Experiment 10.10
Ferroxyl indicator (mix 5 cm^3 of 1% phenolphthalein in ethanol solution and 30 cm^3 of 1% potassium hexacyanoferrate(III) solution with 1 litre of 0.10 mol/l sodium chloride solution), Petri dishes, iron nails, pliers, file, iron wool.

Experiment 10.11
Ferroxyl indicator, agar gel, specimen tubes, iron nails, magnesium ribbon, aluminium foil, copper foil, zinc foil, tin foil.

Experiment 10.12
Aluminium, mercury(II) chloride.

Experiment 10.13
Aluminium foil, Petri dish, copper(II) sulphate solution, sodium chloride, potassium chloride, sodium sulphate, copper(II) sulphate, copper(II) nitrate, sulphuric acid, hydrochloric acid.

Experiment 11.1
Crucible and lid, magnesium ribbon, tongs, tripod, pipeclay triangle, balance.

Experiment 11.2
Copper(II) oxide, reduction tube shown in Fig. 11.2 or porcelain boat and reduction tube, spatula, balance. A better reducing atmosphere is provided by including a small piece of *meta fuel* as shown in Fig. 11.2b.

Experiment 11.3
Solutions of potassium iodide (0.5 mol/l, i.e. 83 g/l) and lead(II) nitrate (0.5 mol/l, i.e. 165.5 g/l), six test-tubes in a rack, two burettes, ruler, graph paper.

Experiment 11.4
Magnesium sulphate crystals, crucible, tripod and pipeclay triangle, balance.

Experiment 12.1
Marble chips or pieces of limestone, watch glass, thick wire, clamp and stand, teat pipette, universal indicator, straw.

Experiment 12.2
Plastic juice bottle, stick, calcium oxide, hydrochloric acid, limewater, bricks.

Experiment 12.3
Copper(II) carbonate, zinc carbonate, sodium carbonate, calcium carbonate and other carbonates, two test-tubes with delivery tubes, two test-tubes, limewater, hydrochloric acid, sulphuric acid.

Experiment 12.4
Marble chips, hydrochloric acid, Buchner flask with rubber tubing attached, trough (glass or plastic or plastic food box or empty food container), two gas jars with lids, four boiling tubes, universal indicator, limewater, magnesium ribbon, tongs, splint.

Experiment 12.5
Metal tray (enamel plate, baking tin etc.), heat-proof mat, wood shavings, wash bottle, sodium hydrogencarbonate, conical flask with jet, ignition tube and cotton thread (see Fig. 12.5).

Experiment 12.6
Fabrics (silk, rayon, cotton, wool, nylon, polyester etc.), heat-proof mat, tongs, sodium borate, boric acid, commercial flame retardant, soap powder (e.g. Lux®), soapless detergent (e.g. Persil®).

Experiment 12.7
Shell, pestle and mortar, two test-tubes with delivery tubes, limewater, splint, universal indicator, hydrochloric acid.

Experiment 13.1
Powdered sulphur, 1,1,1-trichloroethane or methylbenzene, dichloromethane, two boiling tubes, two beakers, tripod and gauze, filter funnel and paper.

Experiment 13.2
Powdered sulphur, combustion spoon, three to six boiling tubes with corks, rack, solutions of potassium dichromate(VI) and potassium manganate(VII), apple peel, leaves and flowers.

Experiment 13.3 Parts A and B
Aluminium potassium sulphate (kept from Experiment 1.2), 250 cm^3 beaker, white cotton cloth, sodium carbonate solution, dyes (Congo red is good, eosin, alizarin), tongs, white tile.

Part C
Filter paper, selection of fabrics.

Experiment 13.4
A permanent crease for wool
Sodium hydrogensulphite, woollen material, steam iron, detergent.
Preserving fruit
Fruit, glassware (e.g. bottles, beakers), Saran wrap, aluminium foil, plastic bags, screw-top jars.

Experiment 13.5
Dilute sulphuric acid, brush or glass rod, filter paper, tongs.

Experiment 14.1
Cheese, fish, meat, soda–lime, activated charcoal, apparatus shown in Fig. 14.1.

Experiment 14.2
Dilute sulphuric acid, ammonia solution, glass rod, beaker, evaporating basin, water bath.

Experiment 14.3
Ammonium sulphate (saved from Experiment 14.2), calcium hydroxide, red litmus paper, copper(II) sulphate solution. Have concentrated hydrochloric acid standing in fume cupboard.

Experiment 14.4
Rock phosphate or calcium phosphate(V), bench sulphuric acid, 100 cm^3 beaker, filter funnel and paper, bench ammonia solution, evaporating basin, water bath.

Experiment 14.5
Seeds (e.g. cress), planters, sand, soil, cotton wool, filter funnel and paper, ammonium sulphate (from Experiment 14.2), ammonium phosphate(V) (from Experiment 14.4).

Experiment 14.6
Ammonium chloride, copper foil or gauze, spatula, test-tube, universal indicator paper.

Experiment 15.1
Chlorine water, 1,1,1-trichloroethane or hexane or tetrachloromethane, bromine water, iodine in aqueous potassium iodide, solutions of potassium bromide and potassium iodide.

Experiment 15.2
Teacher demonstration
Take zinc powder (13.0 g) and powdered sulphur (6.5 g) and shake the elements together on a piece of paper. Pile the mixture on a gauze supported on a tripod in a fume cupboard. Heat it gently with a flame until the reaction starts; then switch off the gas. Collect the zinc sulphide formed for class use.
Class
Boiling tube, zinc sulphide, iodine in aqueous potassium iodide.

Experiment 15.3
Teacher or technician
Make up
a 10% solutions of some household bleaches,
b a 5% solution of sodium thiosulphate,

c a 5% solution of starch.
Class
Solutions **a**, **b** and **c**, three measuring cylinders, burette, conical flask.

Experiment 16.1
Crude oil, rocksil, thermometer, side-arm tube with delivery tube (see Fig. 16.1).

Experiment 16.2
Hexane, candle wax, sodium chloride, sodium hydrogencarbonate, margarine, citric acid, glassware shown in Fig. 16.2a, copper(II) oxide, glassware shown in Fig. 16.2b, anhydrous cobalt chloride paper.

Experiment 17.1
Liquid paraffin, rocksil wool, broken pot, glassware shown in Fig. 17.1.

Experiment 17.2
Teacher or technician
Make up bromine water. Weigh 1.12 g of potassium bromate(V) and 12.0 g of potassium bromide, transfer to a 1 litre flask, add distilled water and 40 cm^3 of bench (2 M) hydrochloric acid; then make up to 1 litre with distilled water. This gives 2.00×10^{-2} mol/l Br_2.

Another procedure is to omit the acid from the standard solution, and add 10 cm^3 of bench acid to the oil in the conical flask. This is better in that the concentration of the bromate–bromide solution remains constant, whereas bromine water loses bromine through vaporisation. The second method is less smelly, but it has the disadvantage that the pupil sees a colourless titrant change into a colourless product, passing through a yellow intermediate stage when bromine has been formed but has not reacted. In the first method, the colour change is much more satisfying.
Class
Bromine water, 1,1,1-trichloroethane or hexane, olive oil, margarine, butter, cooking oil etc.

Experiment 17.3
Apple or other fruit, yeast, filter funnel and paper, glassware shown in Fig. 17.2, limewater.

Experiment 17.4
Dehydration
Ethanol, broken pot, apparatus shown in Fig. 17.4, solution of bromine in trichloroethane or another organic solvent or bromine water.
Oxidation: Teacher or technician
Make acidified sodium dichromate solution by dissolving 25 g of sodium dichromate(VI) in 75 cm^3 of water and adding (**with care**) 10 cm^3 of concentrated sulphuric acid.
Oxidation: Class
Boiling tube, acidified sodium dichromate solution.

Experiment 17.5
Rayon
Copper(II) sulphate, sodium hydroxide solution, glass rod, 0.880 ammonia, filter paper, 250 cm^3 beaker, teat pipette.

Casein
Milk, muslin or other filter, formalin (10% aqueous methanal).
Urea–formaldehyde resin
Fume cupboard, urea, 40% methanal (formalin), cotton wool, plastic cup, sawdust, microscope slide, hydrochloric acid.
Nylon: Teacher or technician
Make up solution A (a 5% solution of 1,6-diaminohexane in water) and solution B (a 5% solution of hexanedioyl chloride (or decanedioyl chloride) in 1,1,1-trichloroethane).
Nylon: Class
Small beaker (smaller than 50 cm^3 if possible), glass rod.

Experiment 17.6
Selection of fabrics.

Experiment 18.1
Boiling tube, thermometer, sherbet, sodium hydrogencarbonate, citric acid, ammonium chloride, ammonium nitrate, potassium nitrate, anhydrous copper(II) sulphate, bench hydrochloric acid, bench sodium hydroxide.

Extension
1 mol/l solutions of hydrochloric acid, sodium hydroxide, ethanoic acid, ammonia, potassium hydroxide, nitric acid, ½ mol/l sulphuric acid, two measuring cylinders, 250 cm^3 beaker.

Experiment 18.2
1 mol/l solutions of sodium hydroxide and hydrochloric acid, burette, thermometer, graph paper.

Experiment 18.3
1.0 mol/l solution of copper(II) sulphate, zinc powder, thermometer, polystyrene cup with lid, graph paper.

Experiment 18.4
Apparatus shown in Fig. 18.4, balance, fuels (e.g. ethanol, methanol, kerosene, hexane but **not** petrol).

Experiment 19.1
Marble chips (small, medium and large), balance, hydrochloric acid, apparatus shown in Fig. 19.1.

Experiment 19.2
Magnesium ribbon cut into 3 cm lengths, bench sulphuric acid, 50 cm^3 beaker, graph paper.

Experiment 19.3
Sodium thiosulphate solution (40 g/l), conical flask, 50 cm^3 measuring cylinder, 10 cm^3 measuring cylinder, bench hydrochloric acid.

Experiment 19.4
Sodium thiosulphate solution (8 g/l), apparatus for Experiment 19.3, thermometer.

Experiment 19.5
Solutions of silver nitrate, potassium bromide, sodium chloride, potassium iodide, filter paper, card.

Experiment 19.6
Hydrogen peroxide solution (2 volume), splint, copper(II) oxide, manganese(IV) oxide, potato, liver (or blood).